Déborah de Oliveira · Maksym Ziberov ·
Alisson Rocha Machado
Editors

ABCM Series on Mechanical Sciences and Engineering

12th Brazilian Manufacturing Engineering Congress - COBEF

ABCM Associação Brasileira de Engenharia e Ciências Mecânicas

Springer

Editors
Déborah de Oliveira
Mechanical Engineering Department
University of Brasília - UnB
Brasília, DF, Brazil

Maksym Ziberov
Mechanical Engineering Department
University of Brasília - UnB
Brasília, DF, Brazil

Alisson Rocha Machado
PUCPR
Pontifícia Universidade Católica do Paraná
Curitiba, Brazil

ISSN 2195-4356 ISSN 2195-4364 (electronic)
Lecture Notes in Mechanical Engineering
ISSN 2662-3021 ISSN 2662-303X (electronic)
ABCM Series on Mechanical Sciences and Engineering
ISBN 978-3-031-43554-6 ISBN 978-3-031-43555-3 (eBook)
https://doi.org/10.1007/978-3-031-43555-3

This Springer imprint is published by the registered company Springer Nature Switzerland AG
The registered company address is: Gewerbestrasse 11, 6330 Cham, Switzerland

Paper in this product is recyclable.

Preface

The "Brazilian Manufacturing Engineering Congress - COBEF" is promoted by the Brazilian Society of Mechanical Sciences and Engineering (ABCM) and is an itinerant event that occurs each two years, representative of all the country. The 12th edition occurred from May 10 to12, 2023, and took place in the capital of Brazil: Brasilia, being hosted by the Department of Mechanical Engineering of the University of Brasilia.

Since its first edition, held in Curitiba/PR in 2001, COBEF has already established itself as the most important event in Manufacturing Engineering in Brazil, occurring in different locations in the country, namely Uberlândia/MG (2003), Joinville/SC (2005), Águas de São Pedro/SP (2007), Belo Horizonte/MG (2009), Caxias do Sul/RS (2011), Penedo/RJ (2013), Salvador/BA (2015), Joinville/SC (2017), São Carlos/SP (2019) and Curitiba/PR (online 2021).

The COBEF 2023 was organized by professors from different institutions, having Dr. Déborah de Oliveira (UnB) as Chair, Dr. Maksym Ziberov (UnB) as Co-chair, Dr. Daniel Monteiro Rosa (UnB) as Treasurer, Dr. Adriano Fagali de Souza (UFSC) as responsible for the Scientific Committee, Dr. Alisson Rocha Machado (PUCPR) as links and actions with industries, Dr. Alexandre Mendes Abrão (UFMG) and Dr. Anna Carla Araujo (INSA Toulouse) as Scientific Editors and Dr. Raphael Lima de Paiva (UFPI) and Dr. Aline Gonçalves dos Santos (UFCAT) as Secretaries.

The event had 406 attendees with access to the lectures and technical sessions of the event. People from 17 different states and the Federal District of Brazil and 8 different countries participated in the event, which took place in a hybrid way. The final program of the event included 5 lectures, 1 round table and the presentation of 227 technical works distributed in 10 different areas of knowledge: Additive Manufacturing, Project and Management, Product Engineering, Processes with Material Removal, Computational Vision, Control Systems, Robotics, Sensors and Actuators, Materials Characterization and Processing, Metrology, Tribology, Union and Assembly Processes and Metallurgical Processes.

Papers that were approved and presented in the 12th Brazilian Manufacturing Engineering Congress - COBEF were invited to resubmit and were carefully selected to take part in this book. They were selected proportionally to the number of studies resubmitted to each area and with a thematic that could represent what was presented during the event. In this sense, this book has 22 chapters. Chapters 1 to 3 are from the areas of Metrology, Tribology and Project and Management, respectively. Chapters 4 to 6 are from Product Engineering and 7 to 9 from Materials Characterization and Processing. Following, Chapters 10 to 15 comprise studies about Additive Manufacturing and finally Chapters 16 to 22 from the area of Processes with Material Removal.

With contributions from leading experts in the field, this book is essential reading for anyone looking to stay up to date on the latest trends and developments in manufacturing engineering. The book provides valuable insights and information to academic researchers, practicing engineers and undergraduate and graduate students.

Editors of the ABCM Series on Mechanical Sciences and Engineering—12th Brazilian Manufacturing Engineering Congress - COBEF.

Déborah de Oliveira
Maksym Ziberov
Alisson Rocha Machado

Contents

In-Situ Measurement and Dimensional Error Modeling: A Case Study in a Machine Tool Manufacturer

Amanda Rossi de Oliveira[1], Thiago Cannabrava de Sousa[1], Thiago Rodrigues Campos da Silva[2], and Erik Gustavo Del Conte[1](✉)

[1] Federal University of ABC, Av. dos Estados, 5001, Santo André 09210-580, Brazil
erik.conte@ufabc.edu.br

[2] B.Grob do Brasil, Av. Caminho do Mar, 1811, São Bernardo do Campo 09609-000, Brazil

Abstract. Quality assurance in manufacturing processes is an issue that will require new strategies to follow the evolution guided by Industry 4.0. Thus, considering the data relevance for this new revolution, it is crucial to understand how inspection and measurement techniques can make the most of the process's information to make quality more manageable and avoid rework. Based on data-driven strategies, the present study aimed to improve the dimensional accuracy of clamp levers and grippers manufactured by milling. A case study was stated on the multinational machine tool manufacturer responsible to produce 42 CrMo4 steel clamp levers and grippers. Starting with a root causes mapping, it was possible to identify critical internal operations for dimensional accuracy variations in the products that require some enhancements. Hence, machined clamp levers and grippers' critical dimensions measured with a coordinate measuring machine (CMM) used in the quality department were compared to those measured with an in-situ measurement probe to improve the online equipment accuracy. Regarding the systematic error evaluation for two critical dimensions in the products, this approach found about 0.019 mm and 0.009 mm absolute corrections to enhance the measurement probe method. Also, cutting tool wear experiments contributed to modeling a compensation approach for this phenomenon during machining, according to the number of manufactured components within the tool life. Finally, the proposed human-based data-driven methodologies contribute to defining the first step of measurement and inspection evolutions through the 4.0 Industry structure in an entire manufacturing system.

Keywords: Metrology · Machining · In-situ measurement · Error modeling · Dimensional accuracy · Tool wear

1 Introduction

The rapid development of new technologies experienced in the last decades has brought the conception of the 4.0 Industry. Emerging as a German initiative, 4.0 Industry aims to integrate state-of-the-art technologies to define a resilient production based on cyber-physical systems [1, 2]. Beyond the technological advent that allows the consolidation

D. de Oliveira et al. (Eds.): COBEF 2023, LNME, pp. 1–12, 2024.
https://doi.org/10.1007/978-3-031-43555-3_1

of this concept, data acquisition, interpretation, application, and storage are crucial for the proper functioning of the 4.0 Industry [2].

Data-driven strategies are a central starting point for the 4.0 Industry applications, considering that each operation involved in a production process has intrinsic limitations [3]. Klingenberg, Borges, and Antunes Jr. [2] showed an extensive systematic literature review that reinforced the need to develop skills for data-based value generation in practical implementations of the 4.0 Industry. In this context, all operations involved in a manufacturing system will need to adequate their data collection and analysis capabilities.

Processing data usage can also benefit the metrology and quality tools in the 4.0 Industry. In this sense, it is crucial to understand how quality tools can take advantage of data, like go/no-go gauges, coordinate measuring machines (CMM), and in-situ measurement probes/sensors used to analyze dimensional accuracy regarding the design tolerances [4]. Thus, identifying potential causes of failure and dimensional variations and guaranteeing a robust process control could be more accessible in the 4.0 Industry.

Although in-situ probes can benefit effective production [4, 5], these measurement systems require reliability to fulfill their role in quality management. In this sense, these in-situ measurement probes or other high-complexity inspection techniques can be compared with conventional offline methods, such as CMM, as an alternative for robustness evaluation [6, 7]. Zanini et al. [7] proposed strategies for X-ray computed tomography (CT) inspection of additively manufactured lattices using the substitution method based on references analyzed with a coordinate measuring machine. As a result, it was possible to state the uncertainties related to the CT inspection based on the conventional CMM, bringing new insights into its implementation as a measurement tool [7]. Similarly, other metrology approaches may be compared to CMM according to a structured methodology to improve its effective quality control capability.

This study aimed to develop data-driven strategies to improve the dimensional accuracy of clamp levers and grippers manufactured by milling. Thus, comparing machined clamp levers and grippers' critical dimensions measured with CMM and an in-situ measurement probe (ISM) based on systematic errors could be used to improve the ISM accuracy. Additionally, experimentally estimating the cutting tool wear also contributed to developing a model to offset this phenomena's effect on dimensional variation during machining. Mapping dimensional deviation sources during processing steps facilitates the identification of different operations contributions to the final product dimensional variation using a data-based approach. Thus, the proposed human-based data-driven methodologies favor achieving the first step for the 4.0 Industry implementation, demonstrating how to incorporate data-based decisions in manufacturing processes.

2 Materials and Methods

The case study was developed in a multinational machine tool manufacturer branch located in the metropolitan region of São Paulo, Brazil. The proposed investigation aimed to revise the machining conditions of 42 CrMo4 steel clamp levers and grippers (Fig. 1) to identify strategies to improve these components' dimensional accuracy by avoiding rework.

The produced clamp levers and grippers are applied in tool magazines to hold the mandrels. These components are made in batches with ten components per machining cycle of about 45 min, running along two regular shifts daily. After manufacturing, grippers are heat-treated with induction hardening and returned for jointing with the clamp levers. This assembling operation is conducted manually, and each clamp lever and gripper must have a compatible fitting system to meet the application requirements. However, some processing-induced dimensional errors can affect the fitting during the set assembling, requiring reworking and impacting the cycle time.

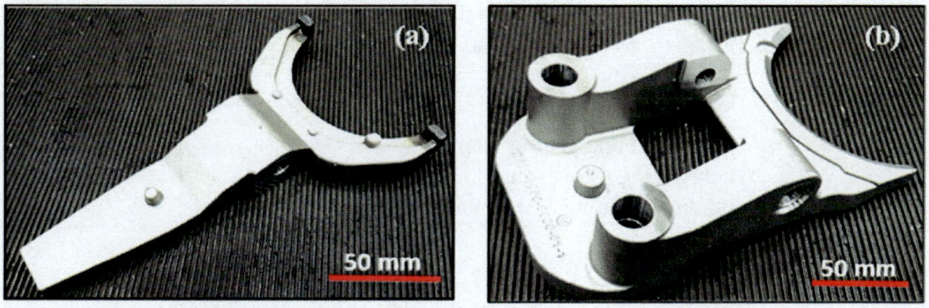

Fig. 1. 42 CrMo4 steel (a) clamp levers and (b) grippers.

Even under rigorous quality control, grippers-clamp lever set manufacturing involves many activities that can introduce dimensional errors. Thus, a root causes mapping could support the main error sources recognition for developing action plans, as further detailed in Sect. 3. As a starting phase for this analysis, a technical visit was carried out to the fabrication cell during a typical working shift to accompany each procedure during the manufacturing process. Then, action plans could be proposed for improving the dimensional accuracy of clamp levers and grippers based on the evaluated activities, as detailed in Sects. 2.1 and 2.2.

2.1 In-Situ Measurement Planning

A horizontal five-axis G550 machining center (GROB-WERKE GmbH & Co. KG, Mindelheim, Germany) performed the milling operations of the clamp levers and grippers. This machine operates with a high-pressure cooling lubricant pumped at about 2.3 MPa [8]. The machine uses a tool magazine to manufacture near-net-shape clamp levers and grippers without multiple setups.

The G550 machining center contained an OMP60 probe (Renishaw, Wotton-under-Edge, England) responsible for on-machine measurement of critical products' dimensions. This device was named ISM in the present study. The ISM measures are commonly considered for manual compensations inputs by the operator if necessary to avoid scrap. However, these adjustments needed a procedure based on the equipment response capabilities in terms of accuracy to prevent introducing dimensional variations in the manufactured components. As a prior work package, the ISM accuracy was evaluated regarding the grippers' critical dimensions in the machining center. So, nine identified grippers had their critical dimensions, named C1 and C2 (Table 1), measured with the ISM using a 175 mm/min feed and later measured with a Global Advantage Hexagon (Stockholm, Sweden) tactile coordinate measuring machine (CMM). Each dimension was evaluated for the milling reference coordinates' positive and negative symmetric sides.

Table 1. Main critical dimensions for the gripper-clamp lever assembly.

Critical dimension	Description	Nominal value (mm)	Tolerances (mm)
C1	Define the value between the gripper's zero position and a Ø12 hole centered in the symmetry axis	140	0/+0.01
C2	Define the diameter of lateral holes in the grippers used in a pin fitting during assembling	Ø10	−0.029/+0.044

The CMM measurements were taken as the calibrated reference due to the expectation of lesser prone-to-error results [4]. Thus, the evaluation of the on-machine method performance used an adaptation of the substitution method regarding CMM calibrations [6, 7]. Therefore, the uncertainty of the ISM (U_{ISM}) used Eq. (1) [6, 7, 9]:

$$U_{ISM} = k_c \sqrt{u_{cal}^2 + u_{dft}^2 + u_{proc}^2 + u_{pc}^2 + u_b^2} \tag{1}$$

where u_{cal} refers to the CMM standard uncertainty; u_{dft} is the uncertainty attributed to workpiece shape variations between the inspection techniques; u_{proc} defines the standard deviation of the reference measurements; u_{pc} considers the uncertainty related to differences between the measured parts in actual working shifts and the ones used for the experiments; u_b is the portion of systematic errors of the ISM measurements, which refers to regular errors that can be managed [3]; and k_c is a constant that defines the coverage factor [6, 7].

The CMM standard uncertainties would be 0.0019 mm for C1 and 0.0015 mm for C2 because of the machine accuracy of $(1.5 + L/333)$ µm, where L corresponds to the length in millimeters. Also, CMM measurements standard deviation (u_{proc}) and the systematic errors attributed to the ISM measurements (u_b) were determined with nine replications [3, 7]. The covered factor k_c was stated as 2, considering the associated confidence interval of 0.95 [6, 7]. Finally, the ISM uncertainty determination disregarded the u_{dft} term, considering that CMM and ISM measurements were conducted just in sequence [7], and the u_{pc} source, taking into account the workpieces used for the measurement evaluation, were real grippers under regular working shifts.

2.2 Tool Wear Evaluation Plan

While machining the most critical geometry in grippers, an arc with beveled edges, the tool wear was also evaluated as a second work package. The experiments were performed with the actual processing conditions, considering the cutting parameters of rotations (n) of 4600 rpm, feed (F) equal to 1380 mm/min, and cutting speed (v_c) of 250 m/min for two tools used for the arc machining. In this scenario, the production rate is of 1.43 parts/minute.

The tools are made up of 16MnCr5 steel. Both milling tools have a similar non-conventional geometry, as illustrated in Fig. 2a, 2b, and 2c. Hence, the tool wear evaluation considered the four critical points of the tool's cutting edge, taking the relative distances between the tool lengths and the preset references located in the cutting edges P1, P2, P3, and P4 (Fig. 2d). The measures were named Z-P_, considering the last digit as the preset identification number. The nominal values of each preset are shown in Table 2.

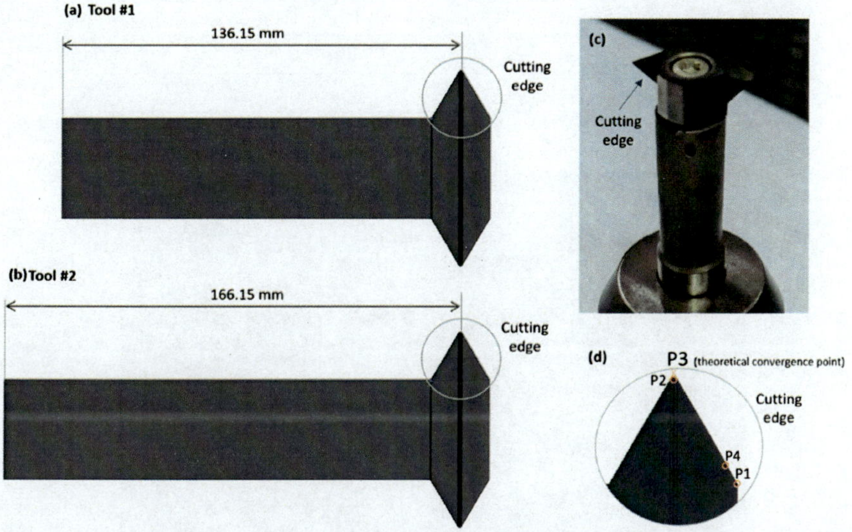

Fig. 2. Illustrative representation of the machining tools (a) #1 and (b) #2 used for tool wear evaluation, as well as (c) the real tool and (d) the preset references in the cutting edges, highlighting the tool wear measurement points P1 to P4.

Table 2. Nominal values of the preset references considered for tool wear evaluation.

Tools	P1 (mm)	P2 (mm)	P3 (mm)	P4 (mm)
#1	136.15	132.30	132.30	135.65
#2	166.15	162.30	162.30	165.65

Three measurement stops equally distributed in a previously known tool life of about 45.5 min were stated for the measurements, as shown in Table 3. For each case the tool was removed from the machining center for inspection in the presetting and measuring equipment venturion 800/10 Zoller GmbH & Co. (Pleidelsheim, Germany), according to the setup shown in Fig. 3. In this system, the tool was positioned in the center of a rotative platform where an industrial camera coupled with a ring light allows the magnification of tool critical parts for a precise inspection. The obtained geometries are synchronized with the pilot computational software, where the quantitative evaluation of the cutting edge presets (Fig. 2d) could be extracted from the processed image of the tool. After each measurement, the tool was returned to the machining center, and processing proceeded until the cycle finished (tool life).

Table 3. Experimental conditions for tool wear evaluation.

Runs	Machining time (min)	Quantity of machined components
#1	10.5	15
#2	28	40
#3	45.5	65

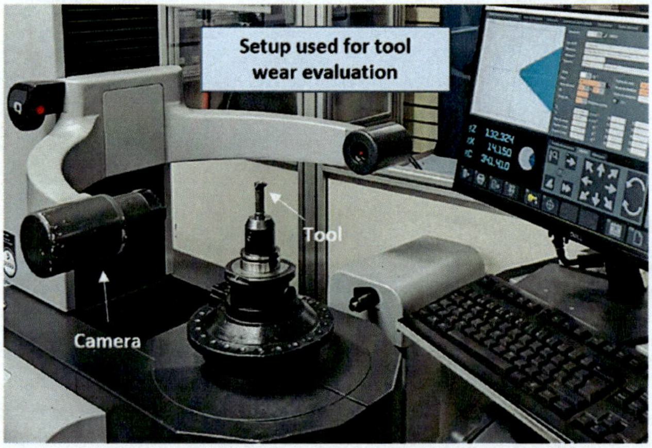

Fig. 3. Equipment used for tool wear inspection.

Following the experimental studies, the dimensional differences found for each preset reference were considered the tool wear when the measurement variation did not influence the expected behavior for this physical phenomenon. Thus, the experimental data could be modeled using linear regression to provide a better in-situ tool wear compensation strategy.

3 Results and Discussion

As a result of the technical visit, an Ishikawa diagram, which allows clustering error sources within a standard classification, could be structured to support the investigations [10]. So, Fig. 4 shows the potential influence factors for dimensional variation in the grippers-clamp lever sets, subdivided by their nature in the processing context. It was covered in the analysis just the internal activities of the manufacturer, which could be improved. Regarding the processing parameters, cutting variables and tool wear can affect the working accuracy of the machine tool, giving rise to geometrical errors [11]. Similarly, the material conditions, like hardness, can provide dimensional variations because of changed material responses during machining [12]. Also, if the stock to leave varies depending on the part, it changes the dimensional accuracy within a working batch. In terms of operations and machines, each activity is susceptible to dimensional variations because of different procedures, occasional mistakes during the setup for different parts, or even noises [13]. The measurement process is another critical source of dimensional variation if there are calibration issues and non-standard statements of the uncertainty relations among the multiple measurement techniques [7]. Finally, environmental conditions, such as temperature, may also induce dimensional deviations because of their potential effects on the material and measurement systems [3].

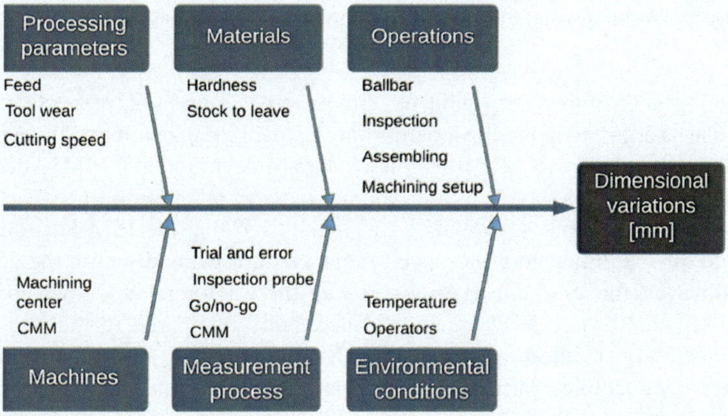

Fig. 4. Ishikawa diagram with the leading potential causes of dimensional variations in the studied machining-based process.

Figure 5 shows the activities in the workflow defined for the case study development. According to the potential causes identified (Fig. 4), experimental tests were planned and conducted during a working shift to be closer to actual processing conditions. Based on the interpretation of the collected data, further detailed in Sects. 2.1 and 2.2, the main root causes that needed improvement were defined. Thus, it was possible to state action plans for solving two issues: in-situ measurement uncertainties and tool wear control during machining (Sects. 2.1 and 2.2). These plans' implementation results are detailed in Sects. 3.1 and 3.2.

3.1 In-Situ Dimensional Measurements

Table 4 shows the statistical metrics obtained for the measurements performed with the ISM and CMM. When comparing both average measurements, systematic errors

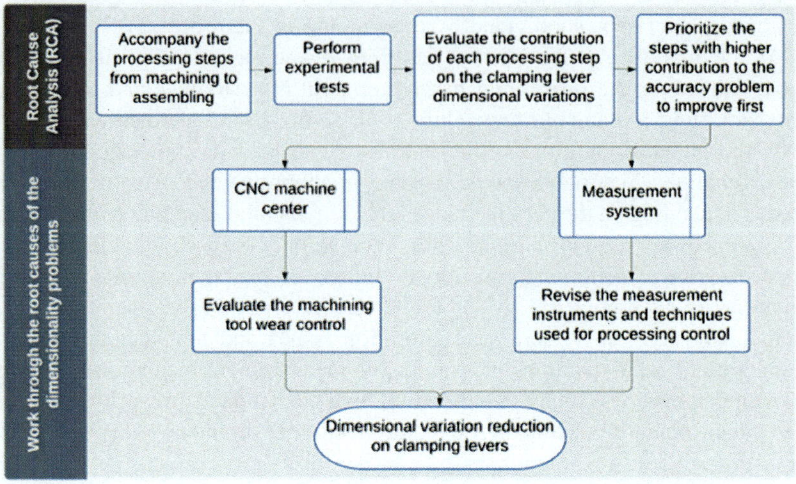

Fig. 5. Workflow used for root cause mapping and operations improvements.

of 0.019 and −0.009 mm were found for dimensions C1 and C2, respectively. These deviations can be adjusted in the ISM to improve the machining results. A lower precision of the ISM may be responsible for the higher standard deviation in CMM. Additionally, the uncertainties of the ISM (U_{ISM}) determined with an adaptation of the substitution method are shown in Table 4. Calculated U_{ISM} showed that the ISM uncertainty was very close to the C2 dimension tolerance (Table 1) but remained within the acceptable interval. However, the ISM uncertainties exceed the C1 tolerance shown in Table 1. Also, it highlighted a high ISM expanded uncertainty compared to the one obtained with the substitution method application for X-ray computed tomography inspection of additively manufactured lattices [7]. Thus, actions should be taken to improve the measurement system.

Table 4. Comparison of the ISM and CMM (reference) measurements for C1 and C2.

Critical dimensions	C1		C2	
	ISM (mm)	CMM (mm)	ISM (mm)	CMM (mm)
Average measurements	140.009	139.990	10.029	10.038
Systematic errors	0.019	-	−0.009	-
Range	0.028	0.037	0.026	0.006
Standard deviation	0.007	0.012	0.001	0.006
U_{ISM}	0.045	-	0.022	-

Figure 6 shows the scatter plots of the experimental ISM measurement data and their estimated systematic errors. Additionally, correlation metrics between ISM measurements and estimated systematic errors were determined to investigate how these errors could be adequately controlled, as shown in Table 5. Finally, C1 and C2 dimensions showed different behaviors between the ISM measurements and estimated errors, evidencing the necessity to elaborate two different routes for variations management.

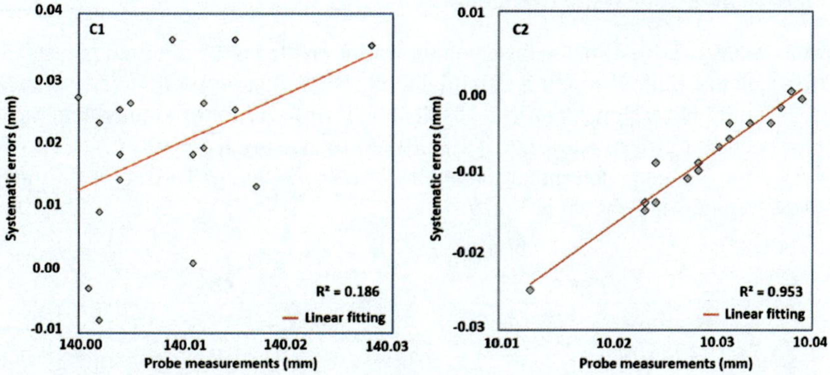

Fig. 6. Scatter plots of ISM measurements and systematic errors for C1 and C2.

Regarding the C1 dimension, based on the Pearson coefficient (r), the coefficient of determination (R^2), and the scatter plot, the lack of relationship between the ISM measurements and respective estimated systematic errors could be stated. In this case, the suggested approach was to use the estimated error value (Table 4) for compensating the in-situ measurements associated with the C1 dimension during machining. This correction could be implemented directly in the NC machining program.

Table 5. Correlation metrics between ISM measurements and estimated systematic errors.

Correlation metrics	C1	C2
Pearson coefficient, *r*	0.431	0.976
p-value	0.074	$4.802 \cdot 10^{-12}$
Coefficient of determination, R^2	0.186	0.953
Adjusted R^2	0.135	0.950

On the other hand, the C2 dimension expressed a high correlation between the ISM measurements and their systematic errors. This conclusion could be supported by the statistics of $r = 0.976$ and $R^2 = 0.953$, shown in Table 4. In this way, it would be possible to implement a linear fitting upon the experimental data for compensating systematic errors generated during the in-situ measurement of the C2 dimension. Equation 2 defines the obtained relation for the adjustment, considering that M_c refers to the measurement compensation and P_m to the ISM measurement. As a result, the established relation could also be implemented in the NC code to improve the dimensional accuracy of the manufactured grippers.

$$M_c = 0.9424 * P_m - 9.4606 \quad (2)$$

3.2 On-machine Tool Wear Compensation

Tool wear values were estimated based on the variations from the nominal presets measured for each machining stop. It is essential to highlight that preset references in which the measures did not follow a feasible physical behavior within the studied interval, for example, decreasing from run #1 to #2 and then increasing from run #2 to #3, were disregarded for the mean determination and tool wear modeling. Thus, plots of the tool wear measurements are shown in Fig. 7.

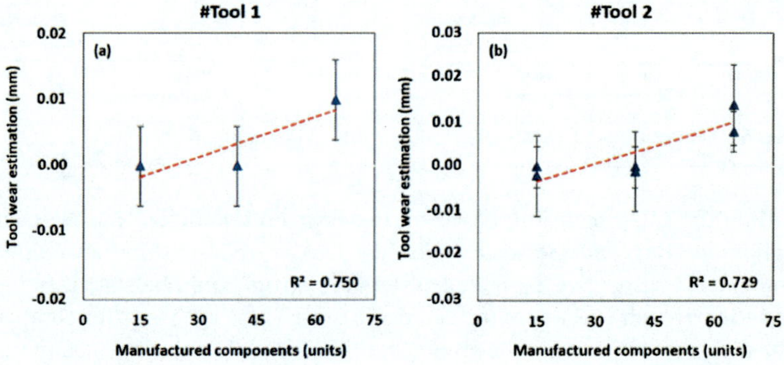

Fig. 7. Tool wear evaluation results for (a) Tool 1 and (b) Tool 2.

Significant tool wear could be identified in the third run, considering values higher than the order of magnitude of tolerances of the critical dimensions (Table 1). This implies that a compensation strategy would be required for dealing with dimensional accuracy. So, the tool wear modeling can be used as the primary dimensional variation control as an on-machine quality tool. Equations 3 and 4 provide the linear fitting models that could be used for tool wear compensation when using tools 1 and 2, respectively, considering the quantity of manufactured components M_{comp}.

$$t_{W_{T1}} = 0.0002 * M_{\mathrm{comp}} - 0.0047 \quad (3)$$

$$t_{W_{T2}} = 0.0003 * M_{comp} - 0.0073 \quad (4)$$

As a next step through the 4.0 Industry, more data could be collected under complementary manufacturing conditions to improve quality control by modeling with machine learning techniques. The proposed data-driven strategies benefit the definition of in-situ monitoring and correction of dimensional accuracy according to actual observations analyzed with statistical tools. So, it would support a transition to a manufacturing structure more suitable for 4.0 Industry pillars, mainly regarding real-time integration and connectivity feasibility [2].

4 Conclusions

Data-driven strategies were proposed to improve the dimensional accuracy of clamp levers and grippers manufactured by milling. Based on experimental studies, it was possible to obtain the systematic errors related to critical dimensions of the grippers using the CMM and ISM comparison and the influence of tool wear behavior through the machining cycle. As a result, two different routes of corrections to be applied directly in the NC program could be proposed to improve the final dimensional accuracy. Regarding the quantitatively determined deviations between ISM and CMM, the C1 dimension could be adjusted based on the observed systematic errors of 0.019 mm. In contrast, the C2 dimension would be better managed by the linear relation dependent on the ISM measurement. Concerning the tool wear, this variation source increased with the manufacturing cycle, reaching values even higher than 0.01 mm after 65 machined components, which is also the value of the C1 tolerance. Thus, linear fitting models were also defined for tool wear compensation to enhance dimensional accuracy. Finally, the proposed approaches take advantage of the data-based decision to improve the in-situ measurement control, highlighting future 4.0 Industry implementation.

Acknowledgments. Special thanks to MSc. Júlia de Andrade Bertazzi from VDI Brasil - Association of Engineers Brazil-Deutschland.

References

1. Lasi, H., Fettke, P., Kemper, H.-G., Feld, T., Hoffmann, M.: Industry 4.0. Bus. Inf. Syst. Eng. **6**(4), 239–242 (2014)
2. Klingenberg, C., Borges, M., Antunes Jr., J.: Industry 4.0 as a data-driven paradigm: a systematic literature review on technologies. J. Manuf. Technol. Manag. **32**(3), 570–592 (2019)
3. Albertazzi Jr., A., Sousa, A.: Fundamentos de Metrologia Científica e Industrial. 1st edn. Manole, São Paulo (2008)
4. Horst, J., Hedberg, T., Barnard Feeney, A.: On-machine measurement use cases and information for machining operations. NIST Adv. Manuf. Ser. **400–1**, 1–69 (2019)
5. Davis, T., Carlson, S., Red, W., Jensen, C., Sipfle, K.: Flexible in-process inspection through direct control. Measurement **39**(1), 57–72 (2006)
6. Schmitt, R., Niggemann, C.: Uncertainty in measurement for x-ray-computed tomography using calibrated work pieces. Meas. Sci. Technol. **21**(5), 1–9 (2010)
7. Zanini, F., Sorgato, M., Savio, E., Carmignato, S.: Dimensional verification of metal additively manufactured lattice structures by X-ray computed tomography: use of a newly developed calibrated artefact to achieve metrological traceability. Addit. Manuf. **47**, 1–8 (2021)
8. Universalmaschine G550. https://www.grobgroup.com/produkte/produktbereiche/universalmaschinen/fraes-bearbeitungszentren/g550/. Accessed 5 Nov 2021

9. International Organization for Standardization: ISO/DIS 15530-3 2009 Geometrical product specifications (GPS) - coordinate measuring machines (CMM): technique for determining the uncertainty of measurement: Part 3. Use of calibrated work pieces or standards (2009)
10. Realyvásquez-Vargas, A., Arredondo-Soto, K., Carrillo-Gutiérrez, T., Ravelo, G.: Applying the Plan-Do-Check-Act (PDCA) cycle to reduce the defects in the manufacturing industry. A case study. Appl. Sci. **8**(11), 1–17 (2018)
11. Fauser, M.: Steuerungstechnische Maßnahmen für die Hochgeschwindigkeits-Bearbeitung: Geometriedatenverarbeitung auf der Basis von Splines und bandbegrenzende Führungsgrößen-Erzeugung. Ph.D. thesis. RWTH Aachen University, Aachen (1997)
12. Bai, Y., Zhao, C., Yang, J., Hong, R., Weng, C., Wang, H.: Microstructure and machinability of selective laser melted high-strength maraging steel with heat treatment. J. Mater. Process. Technol. **288**, 1–18 (2021)
13. Weck, M., Brecher, C.: Werkzeugmaschinen 5: Messtechnische Untersuchung und Beurteilung, Dynamische Stabilität, 7th edn. Springer Vieweg Berlin, Heidelberg (2006)

Investigation of the Effects of Temperature and Contact Pressure on the Friction Coefficient in Uncoated PHS2000 Steel Plates

Bruno Caetano dos Santos Silva[1,2](✉), Luã Fonseca Seixas[3], Luis Fernando Folle[1], Alexandre Ferraz Dourado[1], Tiago Nunes Lima[1], Rodrigo Santiago Coelho[1,3], and Gilmar Ferreira Batalha[2]

[1] SENAI CIMATEC - SENAI Institute of Innovation (ISI) for Forming and Joining of Materials (ISI F&J), Salvador, Brazil
bruno.silva@fieb.org.br

[2] Department of Mechatronics and Mechanical Systems Engineering, Polytechnic School of Engineering, University of Sao Paulo, Sao Paulo, Brazil

[3] Postgraduate Program MPDS/GETEC/MCTI, SENAI CIMATEC, Salvador, Brazil

Abstract. Manganese-boron steels are widely used in hot stamping, resulting in 1500 MPa (PHS 1500) or even 2000 MPa (PHS 2000) tensile strength after the hot stamping process. These steels are most commonly used in industry in the form of AlSi-coated sheets. However, recent developments are being made to enable the application of uncoated steels, resulting in cost reduction, but bringing technological challenges to be overcome. The coefficient of friction (μ) is a fundamental parameter in sheet metal forming and is influenced by several factors, such as: surface roughness, working temperature, lubrication, contact pressure and sliding speed. In hot stamping, it is known that the temperature can have a great influence on μ values for coated plates, but there is little information about uncoated steels. This work aims to investigate the tribological behavior of uncoated PHS 2000 steel plates using the pin-on-disk test. Tests were carried out under three temperatures (800 °C, 600 °C and 400 °C) and contact pressures of 5 and 10 MPa. Results showed that the friction coefficient tends to reduce with increasing temperature at a pressure of 5 MPa, as result of oxide formation on the steel's surface. However, at a pressure of 10 MPa, this reduction was not observed.

Keywords: Friction coefficient · PHS 2000 · pin-on-disk · hot stamping · uncoated

1 Introduction

In the hot stamping process, the metal sheet is heated up to temperatures between 800 °C and 900 °C for at least 5 min, and then it is transferred to the press to be subjected to the simultaneous forming and quenching process in a water-cooled tool [1, 2]. The steels most used in this process are categorized as PHS (Press Hardening Steels), a category of so-called advanced high-strength steels (AHSS) [3]. These materials commonly have

D. de Oliveira et al. (Eds.): COBEF 2023, LNME, pp. 13–21, 2024.
https://doi.org/10.1007/978-3-031-43555-3_2

a metallic coating, usually Al-Si-based, which aims to protect the steel from corrosion and oxidation [1, 2]. The major application of these steels is in the automotive sector, for the manufacture of structural components of several vehicle models, comprising about 38% of the bodywork of modern cars [4, 5]. PHS 2000 steels are susceptible to solid solution hardening of alloying elements such as manganese (Mn) and boron (B) and martensitic transformation. In terms of nomenclature, the number 2000 refers to the approximate maximum limit of mechanical strength (ultimate tensile strength) that this steel can reach [6, 7].

During hot stamping, both the tool and the sheet are subjected to severe friction conditions, which can intensify depending on the complexity of the part geometry. Furthermore, during the hardening step, some thermal shrinkage occurs which promotes relative sliding, under high loads, between the tool and the steel. This combination of temperature and mechanical factors can contribute to excessive tool wear and result in high maintenance costs [3]. Thus, understanding and mapping the tribological behavior of uncoated PHS 2000 steel is essential to improve the applicability of this material in the hot stamping process and its impacts on tool life.

The pin-on-disk friction test is a tribological characterization technique commonly used to estimate the coefficient of friction and wear mechanisms, in addition to being an important method to evaluate the behavior of tribological pairs in terms of friction and wear. This test consists in placing a ring- or disk-shaped specimen in contact with a pin, performing a rotating movement around the axis of the specimen as shown in Fig. 1. In the test, the tangential force and the loading force are measured. Flat or spherical pins can be used. Friction resistance can be determined for different sliding speeds, lubrication conditions and contact pressures. In addition, some pieces of equipment allows to vary the test temperature, placing the sample in a closed and controlled environment [8].

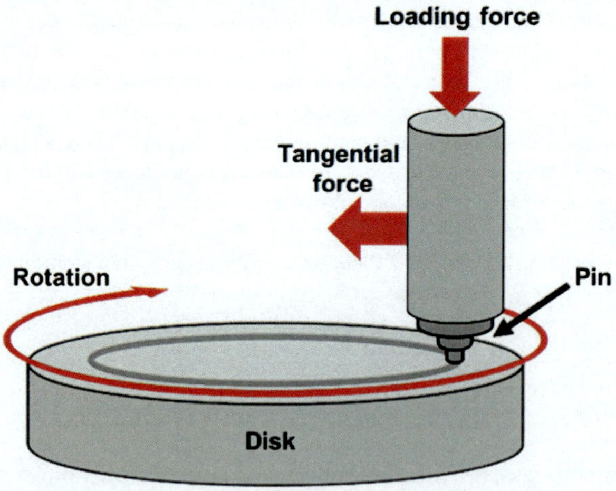

Fig. 1. Schematic representation of a Pin-on-Disk test.

During pin-on-disk testing, the frictional force and sliding distance for the first revolution of the disk are measured using load cells and a computer program to record frictional force values as a function of time. The coefficient of friction "μ" is determined from the values of the normal force of "F_N" and the friction force "F_F", as described in Eq. (1) [8].

$$\mu = \frac{F_F}{F_N} \tag{1}$$

Although there are studies focused on investigating the tribological behavior of uncoated PHS steels, the vast majority evaluates more common steel compositions, such as the PHS 1500 steel. There is still a great lack of evaluation of this phenomenon for the PHS 2000 steel, with superior mechanical properties. Results of investigations related to the tribological behavior of PHS steels indicate, for the most part, that pressure and temperature can significantly affect the value of the coefficient of friction, which varies in the range of 0.4 to 0.8, depending on test conditions [3, 9–11].

Thus, the objective of this study is to evaluate the tribological behavior of PHS 2000 steel plates under high temperature conditions (400, 600 and 800 ºC), considering different conditions of mechanical loading applied to the steel surface (5 and 10 MPa), in order to simulate industrial conditions of the hot stamping process.

2 Methodology

2.1 Samples Preparation

Uncoated PHS 2000 steel plates with 1.2 mm thickness were cut into disks by wire electrical discharge machining (FANUC Robocut Alfa-OiE) of the geometry included machined holes for fixation during tests, as shown in Fig. 2. Dimensions and tolerances were checked by a Coordinate Measuring Machine - CMM (brand Brown&Sharpe, model Scirocco 09-10-07), a caliper (brand Starrett, model 798) and a micrometer (brand Mitutoyo, model 103-129). Cleaning and handling of the specimens were carried out according to the ASTM G115-10 standard, in order to remove substances that could affect results and hinder contact with the tested surfaces. Records of the tested material, test conditions, surface conditions and sample dimensions were made, in order to maintain the traceability of the tests. The surface of the as-received sheet had an average roughness (Ra) of 1.2 μm, measured using a Mitutoyo digital roughness tester, model SJ-210, in accordance with ISO 4287 and ISO 4288 standards. Acetone was used for cleaning. The entire cleaning procedure was performed before heating the samples. Figure 2 shows one of the prepared disks, before testing.

2.2 Hot Friction Tests

Tests were performed using a Brucker pin-on-disk tribometer, model UMT Tribolab, shown in Fig. 3, in accordance with ASTM G115-10 and ASTM G99-17 standards. Aiming to approximate the laboratory experiments to the real hot stamping process, samples were preheated to a temperature of 800 °C, close to the material's austenitization range. Subsequently, cooling was carried out to the target temperature, maintained during the test. Heating was carried out in the tribometer itself, equipped with a coupled heating

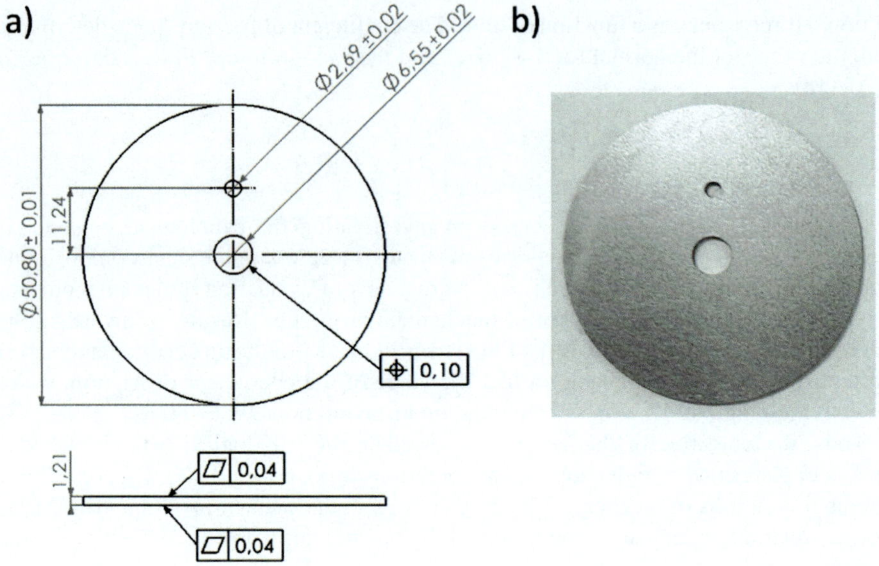

Fig. 2. a) Dimensions (in mm) of the analyzed test disks; b) Appearance of the disk's surface before testing.

furnace, allowing temperature control throughout the test. The use of the coupled furnace for testing allows the heating of samples to occur in a closed and controlled environment, which provides enhanced temperature homogeneity during the test. The parameters used in tests are shown in Table 1. The conditions were set into the equipment's software taking into account the diameter of the pin used and the turning radius, with fixed tangential speed and traveled distance. The equipment applies the pin-disk contact load, controls the disk rotation and reads the tangential and normal forces, which are used to determine the coefficient of friction.

For all specimens, the preheating temperature of 800 °C was maintained for 5 min. The test execution temperatures were 800 °C, 600 °C and 400 °C with established contact pressures of 5 MPa and 10 MPa. As the machine input parameter is force, the loads (98.17 N and 196.34 N, respectively) were calculated based on the pin diameter (5 mm) to obtain the desired pressures. For all test conditions, three repetitions were performed, totaling 18 tests. Two test radii were considered in each disk, as shown in Fig. 4, considering each ray as a repetition. In this way, a total of 9 disks were used. Prior to each test, the pin was ground with 220 μm, 600 μm and 1200 μm grit sandpapers, followed by cleaning with acetone to allow equivalent test conditions in all experiments. After this procedure, the disks showed a surface roughness (Ra) of 0.6 μm.

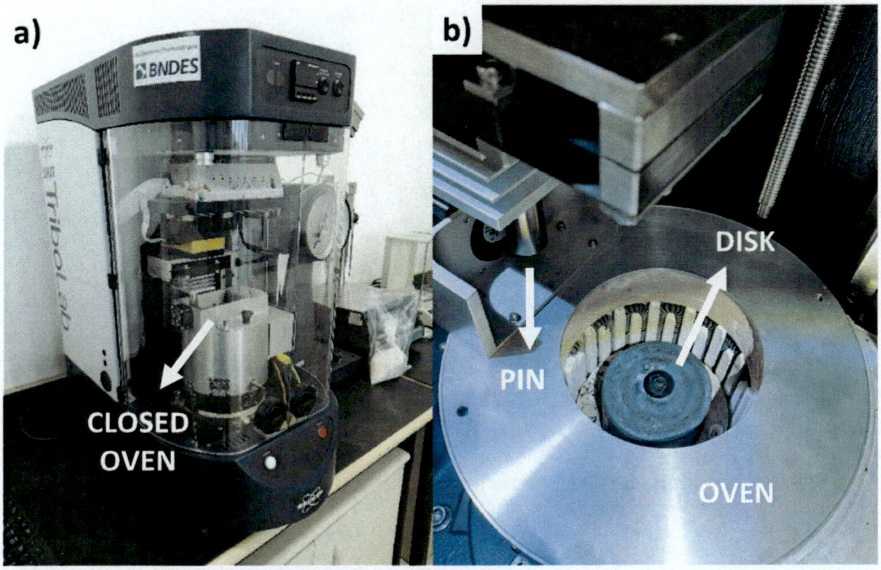

Fig. 3. a) External view of the pin-on-disk test equipment prepared for test execution; b) Internal view of the pin-on-disk test equipment, after performing the test.

Table 1. Parameters applied in the pin-on-disk tests.

Testing Condition	Temperature (°C)	Contact pressure (MPa)	Rotation speed (mm/s)	Traveled distance (mm)
1	400	5	10	600
2	400	10		
3	600	5		
4	600	10		
5	800	5		
6	800	10		

3 Results

Results of the means and standard deviations of the coefficients of friction for each combination of evaluated conditions are shown in Fig. 5. The average value of the coefficient of friction as a function of temperature under the two contact pressure conditions is shown in Fig. 6.

Results suggest a reduction in the friction coefficient with increasing test temperature. The exception was for the tests at 10 MPa, which indicated an increase in the coefficient of friction at a temperature of 800 ºC. This behavior of reduced coefficient of friction with increasing temperature has already been observed in uncoated boron steels, as in the

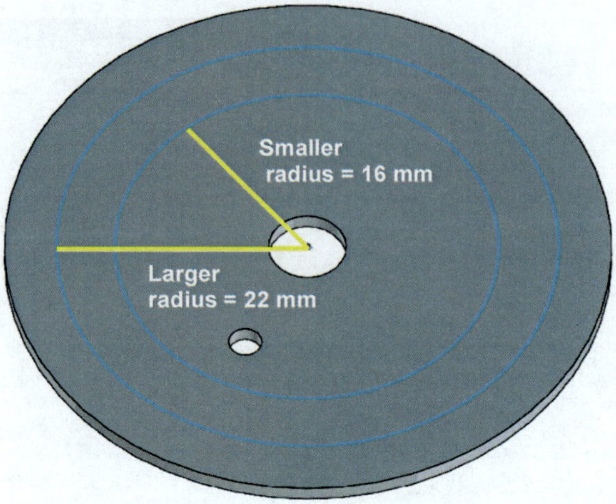

Fig. 4. Analysis radii applied to each test disk, accounting for two tests per disk.

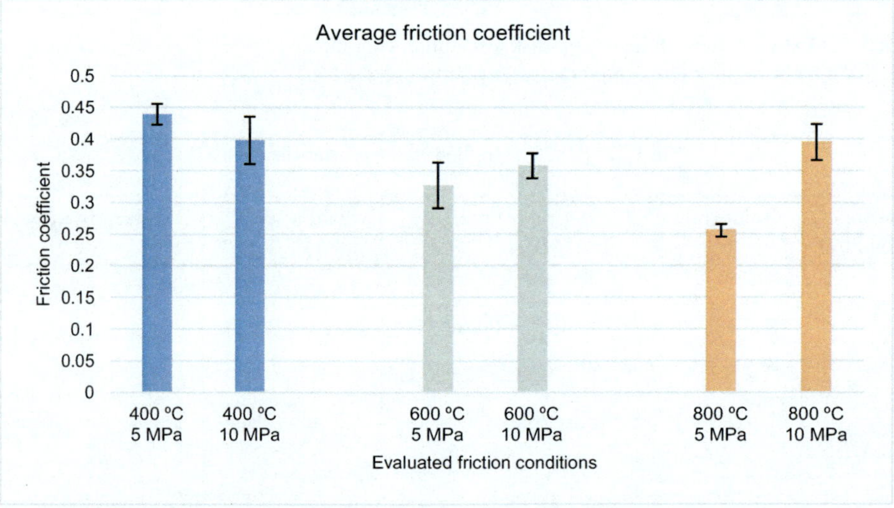

Fig. 5. **Friction coefficients for each combination of evaluated hot conditions.**

work by Hernandez et al. (2014) [9], when evaluating the tribopair boron steel-tool steel in a pin-on-disk test. According to the authors, this can be attributed to the development of oxidized protective layers on the heated disk and pin surfaces, as small changes in temperature and/or load can result in reductions in wear and friction behavior. They elucidate that the tribological behavior of these layers is governed by the composition and type of oxides formed, which depend on factors such as temperature, load, sliding speed and materials in contact. This reduction was also observed in the study by Gracia-Escosa et al. (2017) [3] who evaluated the variation in the coefficient of friction of

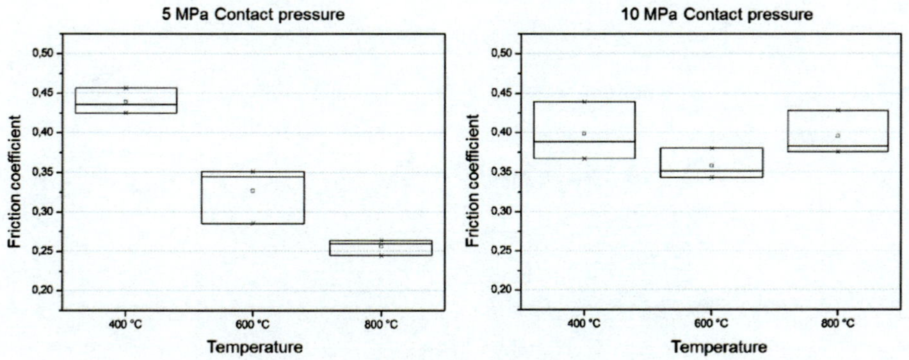

Fig. 6. Friction coefficients by contact pressure.

22MnB5 (PHS 1500) steel at room temperature and at 900 °C. At 900 °C, the uncoated steel showed a reduction in the coefficient of friction.

Figure 7, Fig. 8 and Fig. 9 show representative graphs of the friction coefficient measured in real time during the execution of tests at 400 °C, 600 °C and 800 °C respectively. Results indicate that, with increasing test temperature, there is a tendency to intensify the fluctuation in the frictional behavior of the uncoated PHS 2000, with more expressive variations between peaks and valleys throughout the tests. Furthermore, in most of the conditions tested, an increase in the coefficient of friction was observed at the beginning of the test. Initially, the pin is in contact with the original surface of the disks and, after the first revolution, the pin comes into contact with the surface already worn by the passage of the pin, which can lead to this subsequent reduction in the coefficient of friction after the start of the tests.

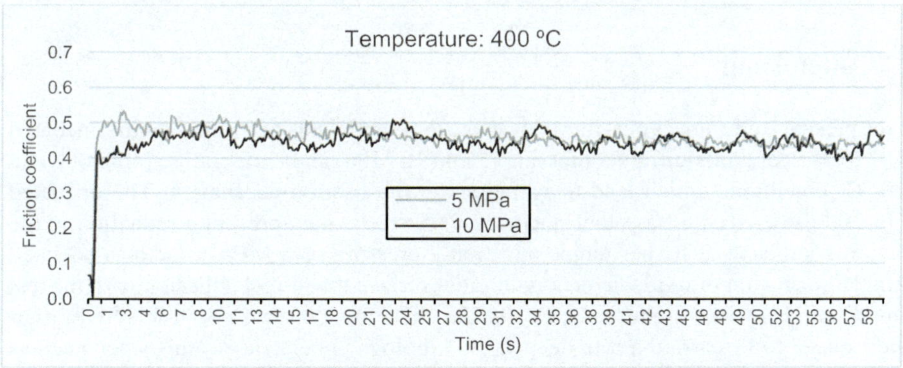

Fig. 7. Graph of friction coefficients over the tests at a temperature of 400 °C.

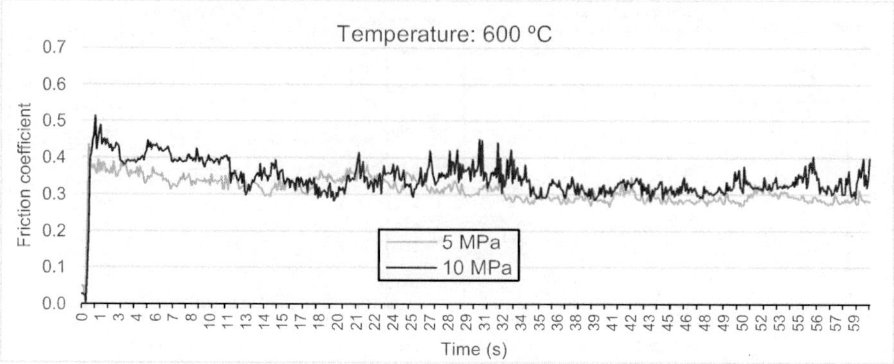

Fig. 8. Graph of friction coefficients over the tests at a temperature of 600 °C.

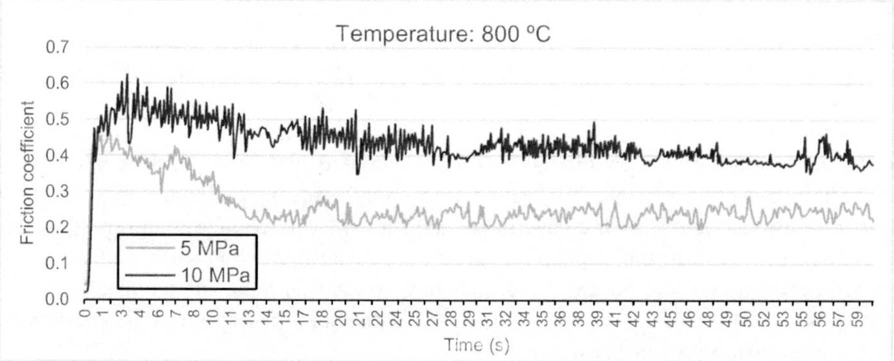

Fig. 9. Graph of friction coefficients over the tests at a temperature of 800 °C.

4 Conclusions

In the present study, it was possible to evaluate the coefficients of friction for the uncoated PHS 2000 steel under different pressure (5 and 10 MPa) and temperature (400, 600 and 800 °C) conditions which lead to variations in the friction coefficient. The uncoated PHS 2000 steel, as well as other uncoated PHS steels, may present a reduction in the friction coefficient at higher temperatures at a pressure of 5 MPa, a factor associated with the formation of oxide layers on the steel surface that affect adhesion with the test pin surface. However, at pressures of 10 MPa, this reduction was not observed. Further studies need to be carried out to deepen the tribological behavior of this steel, such as studies focused on roughness, chemical composition and microstructural analysis.

References

1. Altan, T.: R&D update: hot-stamping boron-alloyed steels for automotive parts - Part I. Stamping J. (2007)

2. Karbasian, K., Tekkaya, A.E.: A review on hot stamping. J. Mater. Process. Technol. **210**(15), 2103–2118 (2010)
3. Gracia-Escosa, E., García, I., Damborenea, J.J., Conde, A.: Friction and wear behaviour of tool steels sliding against 22MnB5 steel. J. Market. Res. **6**(3), 241–250 (2017)
4. Belanger, P.: Steel innovation in hot stamping. In: Great Design in Steel Conference 2016 (2016)
5. Taylor, T., Clough, A.: Critical review of automotive hot-stamped sheet steel from an industrial perspective. Mater. Sci. Technol. **34**(7), 809–861 (2018)
6. Rana, R., Singh, S.B.: Automotive Steels: Design, Metallurgy, Processing, and Applications. Woodhead Publishing (2017)
7. Billur, E.: Hot Stamping of Ultra-High Strength Steels. Springer, Cham (2019). https://doi.org/10.1007/978-3-319-98870-2
8. Trzepiecinski, T., Lemu, H.G.: Recent developments and trends in the friction testing for conventional sheet metal forming and incremental sheet forming. Metals (Basel). **10**(1), 47 (2020)
9. Hernandez, S., Hardell, J., Courbon, C., Winkelmann, H., Prakash, B.: High temperature friction and wear mechanism map for tool steel and boron steel tribopair. Tribol. Mater. Surf. Interfaces **8**(2), 74–84 (2014)
10. Wang, Z., Luo, X.Y., He, W.T., Zhang, Y.S.: Investigation into the tribological behaviors of press hardening steels on the tailored conditions. Sci. China Tech Sci. **58**(1), 97–106 (2015)
11. Hardell, J., Kassfeldt, E., Prakash, B.: Friction and wear behaviour of high strength boron steel at elevated temperatures of up to 800 ºC. Wear **254**(1), 788–799 (2008)

Cleaner Production Principles Applied to the Port Logistics Sector

Évelyn dos Santos Jardim Esteves[1(✉)], Ângelo Márcio Oliveira Sant'Anna[1], Felipe Andrade Torres[2], and Ednildo Andrade Torres[1]

[1] Postgraduate Program in Industrial Engineering, Polytechnic School, Federal University of Bahia, Street Prof. Aristides Novis, 02, Federation, Salvador, BA 40.210-630, Brazil
evelyn.jardim@ufba.com.br

[2] Exact and Technological Sciences Center, Federal University of Recôncavo da Bahia, Street Rui Barbosa, Cruz das Almas, BA 44.380-000, Brazil

Abstract. Much of the electrical energy currently consumed in the industrial sector worldwide comes from non-renewable sources. These, when used, contribute to the formation of greenhouse gases (GHG), atmospheric pollution and global warming. The use of solar energy as an alternative source of power generation has been increasingly employed in companies, reducing operational costs. The geographical location, relief, and climate of the state of Bahia are favorable for solar energy generation. In the first months of 2022, Bahia was responsible for approximately 1/3 of the country's solar energy generation. Company "K", located in the Port of Aratu/BA, operates in the segment of mining, processing and beneficiation of materials derived from magnesium oxide, and performs logistical operations of the raw material for export. In this context, this work is based on the concepts of Cleaner Production (CP), through descriptive and exploratory research, applied to the logistics of the product at the Port of Aratu. Data collection was performed, analysis of indicators that supported the goals intended for the proposed implementation of CP, aiming at environmental management, an opportunity to reduce the consumption of electricity, related to the change of technology, as the objective of this work. Applying the CP concepts in logistics at the Port of Aratu, it is expected that there will be an improvement in processes, providing not only a reduction in electricity consumption but also providing operational and economic gains, in addition to improving the company's image before the market in relation to environmental management and quality.

Keywords: Cleaner Production · Solar Energy · Environmental Management · Energy and Sustainability

1 Introduction

The process of economic development is usually accompanied by an increase in the use of fuels. The search for clean and renewable energy has been a reality among countries and companies since the current energy matrixes in use do not support the demand. It is notorious for the growing concern with the environment, and solar energy gains

D. de Oliveira et al. (Eds.): COBEF 2023, LNME, pp. 22–31, 2024.
https://doi.org/10.1007/978-3-031-43555-3_3

notability, considering that it is part of the clean and renewable energy. Energy sources are one of the main influential vectors in environmental interference, their facilities compromise directly with the dynamics of biotics such as hydro power, petrochemical platforms, geothermal energy and natural gas [1].

The Brazilian electrical system, according to data from the National Agency for Electrical Energy (ANEEL), is driven by generation from various matrixes. The current total installed power in operation in Brazil is 182,974.2 MW, and of this total in operation, more than 83% of the plants are driven by sources considered sustainable, with low greenhouse gas emissions [2].

The use of energy by businesses in various sectors, it's cost, and consumption has been increasing in recent years. Solar energy is among these sources as an option for businesses to generate their own energy. The future of solar energy tends to be favored by increasing global pressures for the use of renewable and clean energy sources and the continuous search for diversification of energy supply sources. This scenario should prevail in the upcoming years and solar energy should be considered as a relevant energy alternative for Brazil [3].

The Earth receives from the sun the energy of about 10 billion Itaipus [4]. Brazil has a favorable location for solar energy generation. It lies geographically in the tropics, which are considered to have a high energy potential. High rates of direct surface solar irradiation occur over much of Northeast Brazil and are the main requirement for the viability of this generation technology.

The change to a vision of sustainability based on the three pillars of ECO-EFFICIENCY: environmental, economic, and social, for a company or a process to be valid being environmentally compatible, economically profitable, and socially fair, implies the adoption of management models that identify the causes of environmental problems to avoid the need for corrective measures, reducing the impacts caused by these on the environment, enabling the definition of alternatives that are economically viable and that effectively contribute to improving the quality of life on Earth [5].

The objective of this work was to evaluate, through the analysis of energy generation information indicators in Brazil, the electric energy consumption of Company "K" in its magnesium oxide loading area, at the port terminal in Aratu/BA, to elaborate a proposal for a CP program, demonstrating the viability of the project for the company and others in the industrial sector in the region.

2 Literature Review

2.1 Brazil's Solar Potential

Knowledge about the potential of the solar resource incident on the surface is essential, but not sufficient to drive the use of this energy source. The variability of the solar resource has impacts on technical aspects of the quality and safety of the electricity system. Thus, in addition to the available potential, reliable information about the variability of the solar resource is essential to support the development of projects for the use of this energy source [6].

According to data from the National Electric Energy Agency [2], 83% of the energy currently consumed in Brazil is generated by sustainable sources, as shown in Fig. 1.

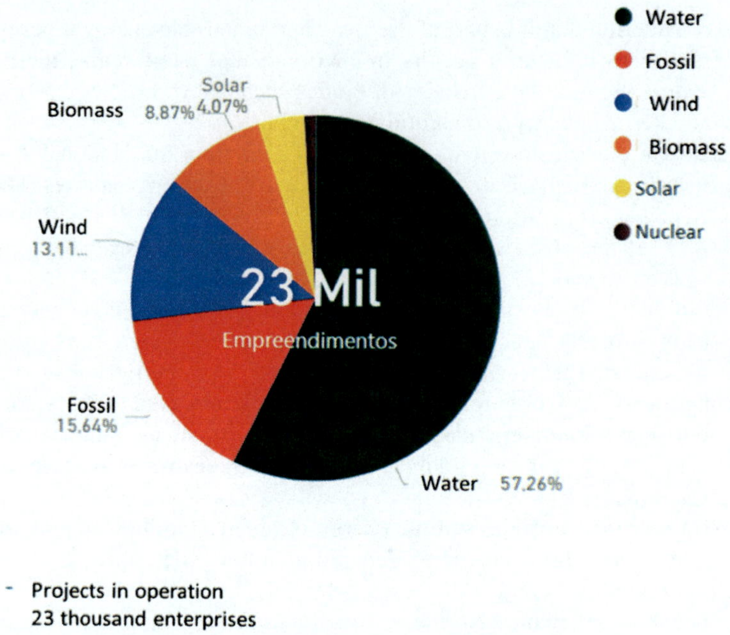

Fig. 1. Electricity matrix by fuel origin [2]

The Northeast of Brazil has the highest rates of solar radiation in the country, which makes it a place of great potential for energy production through the sun's rays. It also presents greater stability in the production of solar energy throughout the year in all seasons [6].

For [7] solar energy stands out in indications of energy sources for diversification of the energy matrix in the country, mainly because of the existing potential little used, but public policies for dissemination of the source could help in the diversification of the energy matrix.

After the ANEEL Normative Resolution No. 482/2012, there was a search for information about the generation of energy by the consumer, since the normative indicates that the user can generate their electricity from renewable sources or qualified cogeneration and even supply the surplus to the distribution network of their locality.

Photovoltaic solar energy is obtained by converting solar radiation into electricity through semiconductor materials. The photovoltaic solar modules are manufactured from silicon (Si) and can consist of monocrystalline, polycrystalline, or amorphous silicon crystals. When sunlight strikes a photovoltaic cell, it produces a small electric current. This current is collected by wires connected to the cell, and transferred to the other components of the system, so the more photovoltaic cells are connected in series or in parallel, the higher the current and voltage produced [6].

2.2 Cleaner Production

The Cleaner Production methodology is the result of a joint effort of UNIDO (United Nations Industrial Development Organization) and UNEP (United Nations Environment Programme), on waste management and energy and material waste [8]. In general, the process of implementing CP follows the flow, as shown in Fig. 2:

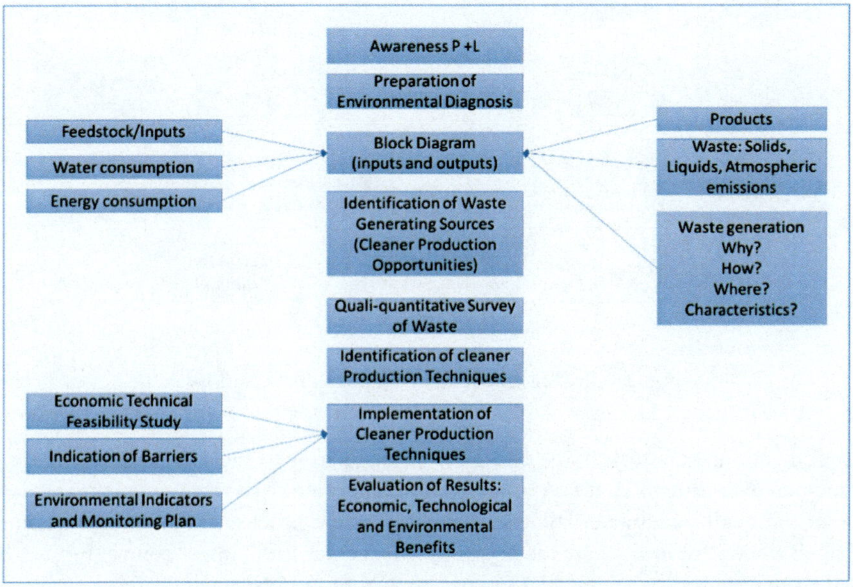

Fig. 2. Flowchart of Cleaner Production implementation - UNIDO/UNEP Methodology [9]

Figure 3 presents the levels of efficiency of application and is a flowchart of the CP, to be implemented by integrating the production processes and the product in companies, to increase productivity, decrease the use of resources, raw materials, water and energy, non-generation, minimization or recycling of waste generated in their activities.

Business strategies, which until then were limited to the economic issue, such as competitiveness, efficiency, profitability, etc., began to introduce their concerns the environmental variable, becoming then vital to incorporate it into production processes [10]. This led to the need for changes in the management and organizational structure, reorienting objectives and goals, and aiming the internalization of environmental issues in management models [11].

2.3 Statistic Process Control

Statistical process control (SPC) is accomplished through control charts, used to monitor the performance of a process from the definition of a control range. Studies developed by Walter Andrew Shewhart, between the 1920s and 1930s, in the United States, at the Bell Telephone Laboratories company resulted from the industrial application, analysis,

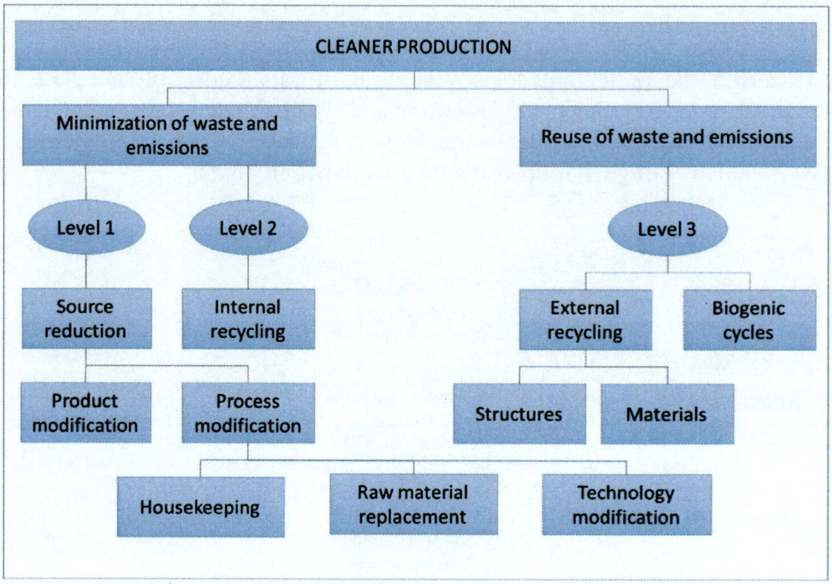

Fig. 3. Cleaner Production Organizational Chart [9]

measurements, and control charts. The control chart is used to analyze trends and patterns that happen over time [12]. It is composed of a center line (LC) that represents the mean value of the quality characteristic corresponding to the process situation under control and a pair of control limits: one located below the center line - lower control limit (LIC) and, the other above the center line - upper control limit (OSL).

Control charts can be classified into two types: for variables and attributes. Con-trol charts for attributes refer to quality characteristics that classify conforming and nonconforming items. Control charts for variables, on the other hand, express quality characteristics on a continuous scale.

[13] argues that SPC enables the increase of process capacity, reduces scrap and rework, and, consequently, the cost of poor quality, providing companies with the basis for improving the quality of products and services.

3 Methodology

The storage facilities of the Port of Aratu, located in the municipality of Candeias/BA, are composed of warehouses, patios, tanks and silos. These installations are leased to private companies, except for the solid bulk yard, which is public and is explored by the Companhia das Docas do Estado da Bahia (CODEBA). Company "K" has a leased warehouse in a public port area, where it operates solid bulk cargoes, resulting from the sintering of magnesium oxide. The solar irradiation index data from the Sérgio Brito Center for Solar and Wind Energy Reference (CRESESB) were used to evaluate the energy potential of the region. Data from the new electric energy generation matrices of the National Electric Energy Agency (ANEEL) were also used.

Cleaner Production is the methodology that proposes strategies to companies to avoid waste of raw materials and energy, waste of any nature, responsible for additional costs to production processes and generate environmental problems.

Statistical process control was used to determine the process variability and statistical analysis of the data, using the R software. The controllable variables were the system current/energy. The energy consumption data was collected through information from the energy provider (COELBA) for the period 2017 and 2021.

4 Results and Discussions

4.1 Analysis of Energy Consumption in Company "K"

Company "K" currently uses the energy distribution system provided by the third largest electric energy distributor in the country in a number of clients and the sixth in the volume of energy distributed, the Companhia de Eletricidade do Estado da Bahia (Coelba).

Using the R software, the electric energy consumption (kWh) between the years 2017 and 2021 was evaluated through the Exponentially Weighted Moving Average (EWMA) control chart, as presented in Fig. 4. This model was employed to estimate and predict the time series of the company's electric energy consumption, the object of this case study.

The Y axis represents the energy consumption (kWh) and the X axis represents the analyzed years, being "1", "13", "25", "37 "and "49" representing respectively January 2017, 2018, 2019, 2020, and 2021.

The quartiles are the separatrices that divide the set into 4 equal parts, to provide a better analysis of the set dispersion, symmetry or asymmetry of the distribution. The system values were: Q1 (first quartile – a value of the set that delimits the 25% smaller values) = 10,214, Q3 (third quartile – a value of the set that delimits the 75% larger values) = 17,735, median = 14,720, maximum = 28,144 and minimum = 4,911 and the average: 14,714, referring to the consumption of electricity (kWh) in Company "K" between 2017 and 2021.

It is also observed that four intervals were identified in which the electric consumption deviated from the control limits (highlighted in red, totaling 19 points), and these occurred between the years 2018 and 2020. These were identified as periods of lower (2018) or higher (2020) cargo handling intensity at the port logistics company (see Fig. 4).

From each observed item, the EWMA Graph enabled the visualization of an exponential forecast (symbolized in the EWMA graph as "+") related to each observation, allowing the study of the situations that fit into a growth or decrease curve of electric energy consumption.

The Graph of individual values (X-bar one chart), Fig. 5, was also prepared, being used for continuous or discrete quantitative variables, as identified in the case study situation, the evaluation of electric power consumption over time.

One can observe a correlation between electric power consumption (kWh) and the months of Aug/20 (x-44) to Dec/20 (x-48), identified in the chart by the yellow sign. During this period in Company "K" there was the closing of the four-months of the

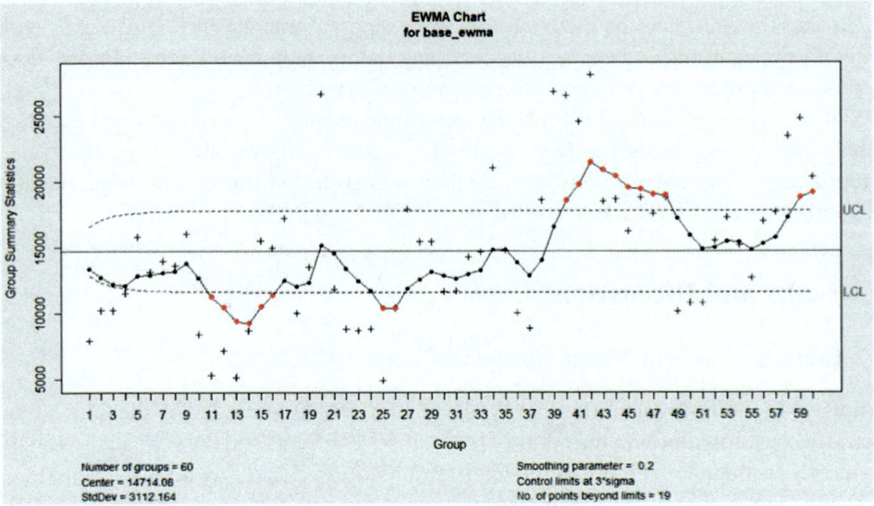

Fig. 4. EWMA chart of electric energy consumption (kWh) at Company "K" between 2017 and 2021 (Authors, 2022)

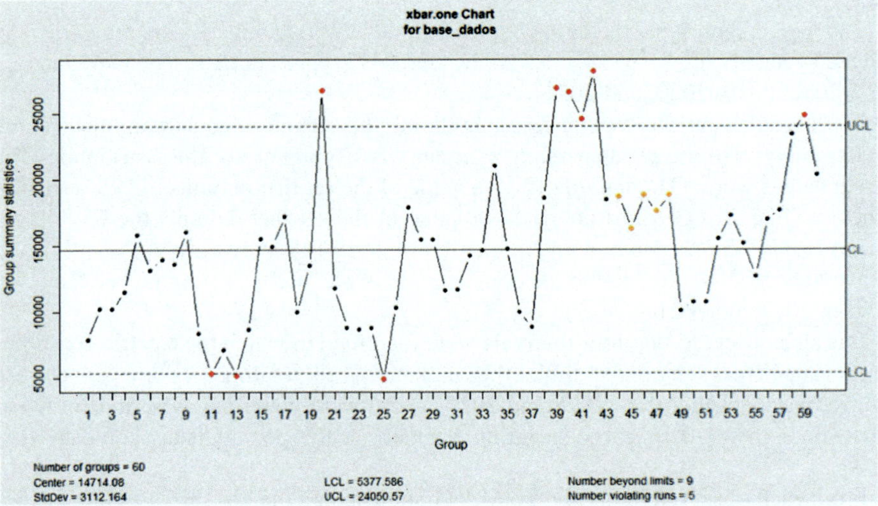

Fig. 5. Moving average chart (X-bar chart) of electric energy consumption (kWh) at Company "K" between 2017 and 2021 (Authors, 2022)

year in logistical operations, right after the resumption of activities after the COVID-19 pandemic.

Analyzing the EWMA control charts (Fig. 4) and the of moving averages (Fig. 5) one can observe a tendency for increase in energy consumption between the months of February and August of the years evaluated. Thus, it is possible to infer that during this interval there was a greater movement of cargo in the referred terminal, either via reception or export, since the reading is performed in a totalized way of the electrical equipment crowded in the structure.

4.2 Opportunity to Implement CP Applied to the Port Logistics Sector

Figure 6 shows that among the new power plants installed in Brazil, there is a growing emphasis on the number of plants that use renewable sources, such as wind and solar energy, whose shares in 2021 would represent 49% and 17%, respectively, of the total installed power.

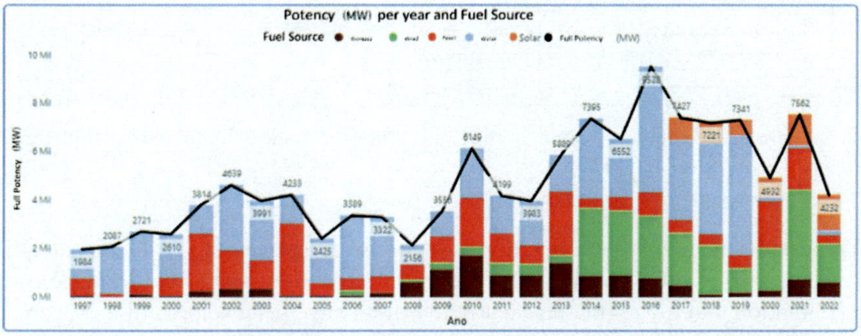

Fig. 6. Installed capacity of new power plants (MW) by year and fuel origin [2]

The use of energy from cleaner energy matrix sources contributes with less impact on nature, reduces electricity losses, optimizes the distribution network and makes consumers aware of more rational energy consumption [14]. The Brazilian territory has reached the mark of 500,000 photovoltaic distributed generation (PVDG) consumer units. According to the Brazilian Association of Photovoltaic Solar Energy (ABSOLAR), adding the wind, thermal, hydraulic and solar sources, the country has more than 511 thousand UCs [15].

Figure 7, shows the solar irradiation index and the monthly average, of the facilities of Company "K" located at the Aratu port terminal, in the municipality of Candeias/BA. Presenting an average monthly solar irradiation of 5.01 kWh/m^2.day.

Thus, through the analyses of this work, it is then suggested the application of the CP strategies related to Process Modification - Technology Change with the implementation of solar tube lighting. This type of technology enables the transfer of natural lighting to the interior of the unit, thus allowing energy savings.

In addition, it is also suggested the installation of a photovoltaic system for the use and conversion of solar energy into electricity, given the possibility of savings with the expense of electricity for Company "K" in the short term. It is also noteworthy that installations of the PVDG type have an estimated average useful life of 20 years, thus allowing the company to take advantage of this investment for a long period. Additionally, it was observed that the unit has a significant roof area available in the cargo storage area (about 5,000 m^2).

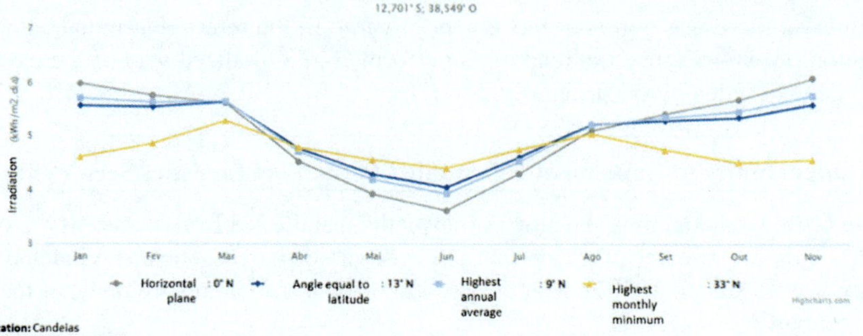

Station: Candeias
City: Candelas, BA - BRASIL
Latitude: 12,701° S
Longitude: 38,549° O
Distância do ponto de ref. (12,783333° S; 38,5° O) :10,6 km

#	Angle	Inclination	Monthly average daily solar irradiation [kWh/m².dia]													
			Jan	Fev	Mar	Abr	Mai	Jun	Jul	Ago	Set	Out	Nov	Dez	Média	Delta
☑	Horizontal plane	0° N	5,98	5,76	5,61	4,51	3,91	3,62	3,87	4,29	5,10	5,41	5,68	6,07	4,98	2,45
☑	Angle equal to latitude	13° N	5,56	5,54	5,64	4,75	4,28	4,04	4,29	4,60	5,21	5,28	5,34	5,59	5,01	1,60
☑	Highest annual average	9° N	5,71	5,63	5,65	4,70	4,18	3,92	4,18	4,52	5,20	5,34	5,46	5,76	5,02	1,83
☑	Highest monthly minimum	33° N	4,60	4,84	5,27	4,77	4,54	4,39	4,63	4,73	5,02	4,72	4,49	4,54	4,71	,88

Fig. 7. Inclined plane solar irradiation (above) and monthly average daily solar irradiation (below) in Cadeias/BA [kWh/m2.day] [16].

The use of clean energy generation systems is recommended by the consumption trend that Company "K" has and also due to the opportunity to take advantage of the solar irradiation that the region has. Currently these have been some of the concerns of companies in different segments with respect to the scarcity of resources, alternative and sustainable sources for the generation of their energy, principles of Cleaner Production that advocate the sustainable use of sources of raw materials, helping to reduce water and electricity consumption.

5 Conclusion

This work sought to describe how the principles of CP can be applied in the sector of transport, storage and loading at the port terminal in Aratu. The geographical location of the enterprise allows the installation of clean energy systems for on-site power generation, being a cost-effective, profitable, and sustainable project.

The analysis of the data, through control charts, referring to the company's electricity consumption, as well as the identified potential for solar power generation and the rise of new energy matrices were necessary to present a proposal for a PVDG system. Studies and proposals aimed at sustainability, quality, and environmental management are increasing of interest to companies seeking to reduce the consumption of resources since they are following the new energy matrices on the rise in the country, such as public policies, government projects, Normative (e.g. No. 482/2012) that encourage the generation of energy itself.

References

1. Silva, C.H.S.T., et al.: New solar energy technology trends. Braz. J. Dev. Curitiba **8**(8), 55730–55750 (2022)
2. ANEEL: Agência Nacional de Energia Elétrica. Sistema de informações de geração SIGA. Feveiro (2023)
3. Marques, F.M.R.: Perspectivas para a Energia Solar no Brasil. Disponível em. https://crasp.gov.br/wp/wp-content/uploads/Artigo-Perspectivas-para-a-energia-solar-no-Brasil.pdf. Acesso: fev. 2023
4. Stensmann, B.H.W.: Energia Emitida pelo Sol. Disponível em. https://ppgenfis.if.ufrgs.br/mef008/mef008_02/Berenice/aula4.html. Acesso fev. 2023
5. Sadler, B.: Environmental Assessment in a Changing World: Evaluating Practice to Improve Performance. International Study of the Effectiveness of Environmental Assessment, Ottawa (1994)
6. Pereira, E.B., Martins, F.R., de Abreu, S.L., Rüther, R.: Atlas brasileiro de energia solar. 2 edn. São José dos Campos: INPE (2017)
7. Rosa, A.R.O., Gasparin, F.P.: Panorama da Energia Solar Fotovoltaica no Brasil. Revista Brasileira de Energia Solar. Ano **VII**(2) (2016)
8. Kiperstok, A., et al.: Prevenção da poluição. Brasília: SENAI/DN, 290 p. (2002)
9. CNTL: Centro Nacional de Tecnologias Limpas Implementação de Programas de Produção mais Limpa. UNIDO/UNEP/CNTL/SENAI. Porto Alegre/RS, 42 p. (2000)
10. Silva, A.L.E., Moraes, J.A.R., Machado, E.L.: Proposta de produção mais limpa voltada às práticas de ecodesign e logística reversa. Eng. Sanit. Ambient **20**(1) (2015). https://doi.org/10.1590/S1413-41522015020000087843
11. Piotto, Z.C.: Eco-eficiência na Indústria de Celulose e Papel - Estudo de Caso. Tese (Doutorado) – Escola Politécnica da Universidade de São Paulo, São Paulo (2003)
12. Montgomery, D.C.: Introduction to Statistical Quality Control, 7th edn. (2016)
13. Ribeiro, J.L.D., Cate, C.S.T.: Cartas de Controle para Variáveis, Cartas de Controle para Atributos, Função de Perda Quadrática, Análise de Sistemas de Medição. Série monográfica Qualidade - Controle Estatístico do Processo - Porto Alegre: FEENG/UFRGS, 172p. (2012). (Série Monográfica Qualidade) ISBN 85-88085-10-0 Universidade Federal do Rio Grande do Sul. Escola de Engenharia. Programa de Pós-Graduação em Engenharia de Produção. III. Título. IV. Série CDU519.2. 2012
14. Cardoso, D.S., Locatelli, P.S., Ramalho, W., Asgary, N.: Distributed generation of photovoltaic solar energy: impacts of Aneel's new regulation proposal on investment attractiveness. Rev. Adm. UFSM **14**(2) (2021). https://doi.org/10.5902/1983465961993
15. Portalsolar: No setor de geração distribuída fotovoltaica, Brasil alcança marca de 500 mil unidades consumidoras - Portal Solar - Tudo sobre Energia Solar Fotovoltaica. Disponível em. https://www.portalsolar.com.br/blog-solar/geracao-distribuida/no-setor-de-geracao-distribuida-fotovoltaica-brasil. Acesso: nov. 2022.
16. CRESESB: Centro de Referência para Energia Solar e Eólica Sérgio De Salvo Brito; Energia Solar: Princípios e Aplicações. Tutorial Solar. Disponível em. http://www.cresesb.cepel.br/index.php#data. Acesso em setembro 2022
17. Radonjic, G., Tominc, P.: The role of environmental management system on introduction of new technologies in the metal and chemical/paper/plastics industries. J. Clean. Prod. **15**, 1482–1493 (2007)
18. Severo, E.A., Olea, P.M.: Metodologias de Produção mais Limpa: Um Estudo de Caso no Pólo Metal-Mecânico da Serra Gaúcha. Revista Ingepro – Inovação, Gestão e Produção **2**(7), 73–81 (2010)

A Model Based on Non-linear Regressions to Predict Aluminum Injection Moulds Lifespan

Evandro Menezes de Souza Amarante[1],
Victor Gabriel Sousa Fagundes dos Santos[1],
Pedro Guilherme Carvalho de Souza Marconi[1](✉),
Cristiano Vasconcellos Ferreira[2], Valter Estevão Beal[3],
and Armando Sá Ribeiro Junior[1]

[1] Mechatronics Graduate Program, Federal University of Bahia, Salvador 40210-630, Brazil
{evandro.amarante,victor.sousa,pedro.marconi,asrj}@ufba.br
[2] Federal University of Santa Catarina, Florianópolis, SC 88010-970, Brasil
cristiano.v.ferreira@ufsc.br
[3] SENAI CIMATEC, Av. Orlando Gomes, 1845 - Piatã, Salvador, BA 41650-010, Brasil
valtereb@fieb.org.br

Abstract. Traditionally, thermoplastic injection moulds are made of steel due to its good thermal stability, high mechanical resistance and the possibility of easily obtaining the desired surface finish. Despite these advantages, steel has low thermal conductivity when compared to other materials, such as aluminium, thus requiring longer periods for the part to cool before demoulding it and, consequently, generating longer operating cycles. On the other hand, the use of aluminium in the manufacture of tools used in the injection process is still limited due to its low mechanical strength (when compared to steel), leading to distrust as to the number of parts that can be injected and, therefore, as to their lifespan. The lack of data regarding the frequency of maintenance that must be performed on the aluminium mould to ensure the quality of the injected parts corroborates this scenario of uncertainty, since this information is essential to guarantee the viability of these moulds for high production of plastic parts. This work proposes a mathematical model based on exponential regressions in order to estimate the maximum number of cycles that can be performed before the need for preventive maintenance, and to evaluate the lifespan of aluminium moulds. In order to achieve this, the model was based on the number of cycles related to preventive maintenance, as well as the mechanical strength and hardness of various mould materials.

Keywords: Injection moulding process · Aluminum injection mould · mould lifespan · exponential regressions

D. de Oliveira et al. (Eds.): COBEF 2023, LNME, pp. 32–45, 2024.
https://doi.org/10.1007/978-3-031-43555-3_4

1 Introduction

In the area of thermoplastic injection moulding, several process parameters must be considered to ensure that plastic parts with different degrees of complexity can be produced with the desired quality. Thus, in order to achieve the established goals, it is necessary to correctly select the parameters used, including the injection mould.

Steel is widely used in moulds for injection process due to its high mechanical resistance, good thermal stability and the possibility of easy finishing that it presents. However, its low conductivity (when compared to other materials, such as aluminium) leads to relatively long operating cycles and, consequently, reduces productivity. In this perspective, the use of aluminium alloys guarantees a good and fast heat distribution and dissipation, which results in reduced cycle times and can lead to higher productivity and better quality of produced parts [1].

On the other hand, according to Arieta Filho *et al.* [2], although aluminium has better thermal properties and superior machinability when compared to steel, it has lower mechanical property values, leading to concerns about the number of safe operating cycles and its lifespan. Regarding this, it is important to emphasize that, during the injection moulding process, the moulds are subjected to high temperatures gradient due to the contact with the softened material and cooling fluids, and exposed to the action of external loads. Consequently, these cyclical and thermal efforts, and physical-chemical attacks can cause tooling wear.

These factors, associated with the reduced use of aluminium moulds for high production, corroborate not only the lack of studies related to the performance of these moulds, but also the scarcity of data regarding the periodic maintenance that must be carried out to guarantee the final quality of the injected parts and the longevity of the mould. Such information is essential both for increasing the productivity of the process and for creating a safe scenario for the use of aluminium.

It is worth mentioning, however, that this scenario is becoming more optimistic since there are now international automakers using aluminium for high production .2. There have also been advances in parameter control and the methodology applied to preventive maintenance presented in [3], which, in turn, constituted an important object of study.

Menges and Mohren [1] point out that parameters such as the polymer used in the process (with or without charge), the processing conditions (injection pressure, temperature), and moulding geometry foster a set of variables that need to be investigated about injection mould integrity. As the mould shows surface wear, for example, its closure can be compromised, leading to material leakage through the parting line and, consequently, the appearance of burrs in the injected parts. This type of irregularity must be avoided and is an indication of maintenance breaks for injection moulds, which lead to greater tooling unavailability and wear, and should be carefully examined. As the use of aluminium moulds is not widespread, there is a lack of information regarding its lifespan and autonomy, therefore it is necessary to develop models that estimate the maintenance cycles and longevity of this type of tool.

From this perspective, aiming the evaluation of the lifespan and frequency of maintenance in aluminium moulds, this work proposes a mathematical model based on exponential regressions to predict the number of cycles supported by moulds made of this material. The model was based on numbers of cycles related to preventive maintenance, as well as values of mechanical strength and hardness of different materials.

Mathematical models based on exponential regressions were used to predict the number of cycles and mould lifetime. The motivation for using these models comes from the simplicity presented, in addition to the good accuracy related to their mathematical pattern. It should be noted that exponential models present fast practical and computational implementation [4]. This approach is widely used in different areas of activity, highlighting its great applicability in the financial sector. However, there are also studies associated with underexplored areas, such as works related to epidemics [5] and on indicators related to the integrity of bearings [6].

2 Method and Materials

Considering the uncertainty of information about the use of aluminium in moulds for thermoplastic injection, a mathematical model based on exponential regressions was developed to estimate the maximum number of cycles supported by a mould before preventive maintenance is required. Table 1 shows the list of materials used in the manufacture of injection moulds and whose properties (hardness and average resistance) served as a reference for adjusting the curve.

Table 1. Mechanical properties of steels (Table of N°. of injections, 2022).

Steel	Resistance (kgf/mm^2)	Average Resistance (kgf/mm^2)	Brinell Hardness (HB)	Average Brinell Hardness (HB)
1045	61/67	64	169/187	178
1045 Ecofast	71/78	74,5	197/207	202
Protomold 1050	75/82	78,5	207/229	218
2311	91/100	95,5	269/293	281
2311 st	97/106	101,5	285/302	293,5
PMo Revolution (P20M)(1)	97/110	103	269/321	295
Ecoplast (P20)(1)	97/110	103	269/321	295
2738 HPM 32	101/110	105,5	293/311	302
SP 300	101/110	105,5	293/311	302
PMo Revolution (P20M) ST(2)	100/112	106	280/311	295,5
Ecoplast (P20) ST(2)	97/112	106	280/311	295,5

(continued)

Table 1. (*continued*)

Steel	Resistance (kgf/mm^2)	Average Resistance (kgf/mm^2)	Brinell Hardness (HB)	Average Brinell Hardness (HB)
2738 HPM 35	110/120	115	311/340	325,5
PMo Revolution (P20M) ST(3)	119/132	125,5	311/375	343
Ecoplast (P20) ST(3)	119/135	127	331/375	353
2738 HPM 40	123/132	127,5	363/388	375,5
2711	123/132	127,5	363/388	375,5
SP 400	123/132	127,5	363/388	375,5
2738 HPM 45	136/146	141	401/429	830
ESK 2344 Temp	149/158	153,5	495/514	504,5

Tables 2 and 3 contain data referring to the number of cycles supported by the moulds until a certain preventive maintenance is required. In this study, samples collected between the first and eleventh maintenance were used for mould projects working under identical conditions (type of polymer and processing) and whose distinction lies only in the type of material (steel).

Table 2. Number of cycles until the fifth mould maintenance (Table of No. of injections, 2022)

Steels	Injections - Cycles				
	Maintenance 1	Maintenance 2	Maintenance 3	Maintenance 4	Maintenance 5
1045	10000	25000	32500	40000	50000
1045 Ecofast	13846	34615	45000	55385	69231
Protomold 1050	18462	46154	60000	73846	92308
2311	53846	134615	175000	215385	269231
2311 st	76923	192308	250000	307692	384615
PMo Revolution (P20M)(1)	76923	192308	250000	307692	384615
Ecoplast (P20)(1)	92308	211538	300000	338462	423077

(*continued*)

Table 2. (*continued*)

Steels	Injections - Cycles				
	Maintenance 1	Maintenance 2	Maintenance 3	Maintenance 4	Maintenance 5
2738 HPM 32	92308	211538	300000	338462	423077
SP 300	92308	211538	300000	338462	423077
PMo Revolution (P20M) ST(2)	92308	211538	300000	338462	423077
Ecoplast (P20) ST(2)	92308	211538	300000	338462	423077
2738 HPM 35	365384	365384	475000	584615	730769
PMo Revolution (P20M) ST(3)	750000	750000	750000	923076	1153846
Ecoplast (P20) ST(3)	750000	750000	750000	923076	1153846
2738 HPM 40	750000	750000	750000	923076	1153846
2711	750000	750000	750000	923076	1153846
SP 400	750000	750000	750000	923076	1153846
2738 HPM 45	1250000	1250000	1250000	1538461	1923076
ESK 2344 Temp	3000000	3000000	3000000	33692307	4615384

As can be seen in the previous tables, the present data refer only to steels. However, the model to be developed is mainly a function of the mechanical properties of the materials (i.e. strength and average hardness), so that it can be used for materials other than steel, as long as the properties are within the range taken as a reference for the prediction. In addition, Startks [7] stresses that constructive differences in mould design, surface and thermal treatments and repairs must be considered in the qualitative analysis of results.

Once the data set for analysis was established, the following procedures were performed: (1) data analysis and selection, (2) data pre-processing and (3) model training and validation.

Table 3. Number of cycles from the sixth to the eleventh mould maintenance (Table of No. of injections, 2022).

Steels	Injections - Cycles					
	Maintenance 6	Maintenance 7	Maintenance 8	Maintenance 9	Maintenance 10	Maintenance 11
1045	60000	70000	80000	90000	100000	118182
1045 Ecofast	83077	96923	110769	124615	138462	163636
Protomold 1050	110769	129231	147692	166154	184615	218182
2311	323077	376923	430769	484615	538462	636364
2311 st	461538	538462	615385	692308	769231	909091
PMo Revolution (P20M)(1)	461538	538462	615385	692308	769231	909091
Ecoplast (P20)(1)	507692	592308	676923	761538	846154	1000000
2738 HPM 32	507692	592308	676923	761538	846154	1000000
SP 300	507692	592308	676923	761538	846154	1000000
PMo Revolution (P20M) ST(2)	507692	592308	676923	761538	846154	1000000
Ecoplast (P20) ST(2)	507692	592308	676923	761538	846154	1000000
2738 HPM 35	876923	1023077	1169231	1315385	1461538	1727273
PMo Revolution (P20M) ST(3)	1384615	1615384	1846253	2076923	2307692	2727272
Ecoplast (P20) ST(3)	1384615	1615384	1846253	2076923	2307692	2727272
2738 HPM 40	1384615	1615384	1846253	2076923	2307692	2727272
2711	1384615	1615384	1846253	2076923	2307692	2727272
SP 400	1384615	1615384	1846253	2076923	2307692	2727272
2738 HPM 45	2307692	2692307	3076923	3461538	3846153	4545454
ESK 2344 Temp	5538461	6461538	7384615	8307692	9230769	10909090

2.1 Data Analysis and Selection

Initially, in order to verify if there was a correlation between the data referring to the number of cycles related to each maintenance and the variables average hardness (HB) and average resistance (kgf/mm^2), two scatter plots were generated, as can be seen in Fig. 1.

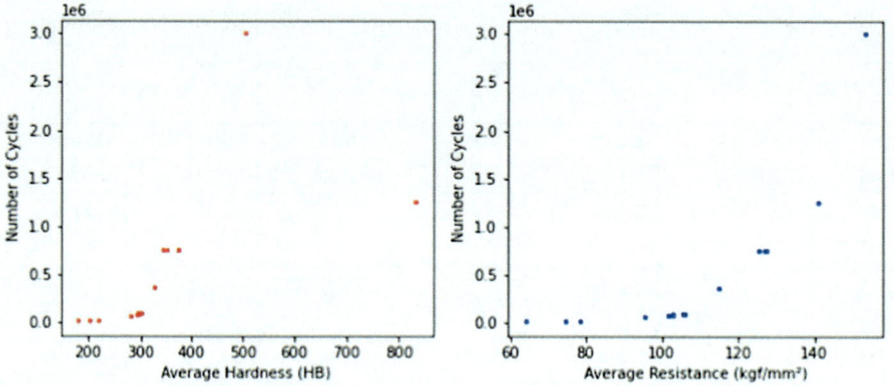

(a) (b)

Fig. 1. Number of cycles to first maintenance: (a) based on average hardness (HB) and (b) based on average strength (kgf/mm2) (The authors, 2022).

The distribution of the points on the plane indicates a tendency in the data behaviour. As presented in Table 1, there are materials that have the same value for mechanical properties and, consequently, identical values for the number of cycles (Fig. 1). Thus, in order to clarify the behaviour of the data, it was necessary to remove the following steels: 2311 st, Ecoplast (P20)(1), 2738 HPM 32, SP 300, Ecoplast (P20) ST(2), 2738 HPM 35, PMo Revolution (P20M) ST(3), Ecoplast (P20) ST(3), 2738 HPM 40, 2711, SP 400, 2738 HPM 45.

Box diagrams were also used for the number of cycles, which are, according to Williamson *et al.* [8], visual tools for observing outliers, distribution and symmetry in a data dispersion, as can be seen in Fig. 2.

According to the graph in Fig. 2, there are cases in which the number of cycles deviates from the general behaviour present in the data set. Considering that parametric regression algorithms are affected by the presence of outliers [9], all cases in which the number of cycles was equal to or greater than 106 were discarded.

Consequently, the dataset included data from the following materials: 1045, 1045 Ecofast, Protomold 1050, 2311, PMo Revolution (P20M)(1) and PMo Revolution (P20M) ST(2). For this set, two new scatter plots of the number of cycles until the first maintenance were generated as a function of the variables hardness and average resistance, as can be seen in Fig. 3.

The tendency present in the dispersions of Fig. 3 was repeated for the other maintenance, so that, once the number of maintenance operations was also considered as an

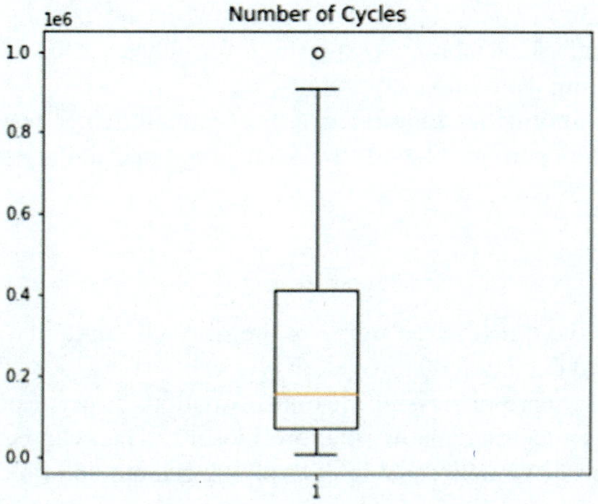

Fig. 2. Number of cycles box diagram

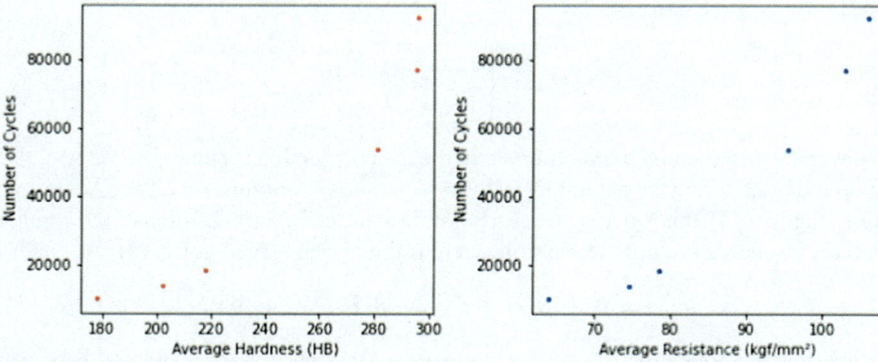

(a) (b)

Fig. 3. Number of cycles until the first maintenance after treatment: (a) based on average hardness (HB) and (b) based on average strength.

independent variable, it was possible to develop a model for predicting the number of cycles referring to all the maintenance operations considered.

Considering the similarity between the dispersions in Fig. 3, which is repeated for the other maintenance, the exponential regression was chosen as a function to be used in predicting the number of cycles.

2.2 Data Pre-Processing

According to Mustaffa and Yusof [10], pre-processing is the step in which the input and/or output variables of a learning model are transformed in order to obtain an improvement in performance. Halim *et al.* [11] mentions that among the existing tools, normalization techniques stand out, which are responsible for changing the values of numerical attributes and inserting them into a common scale.

One example of a normalization strategy is the StandardScaler, responsible for adapting any range of values to an interval of zero average and unitary standard deviation Eq. (1)

$$z = \frac{x - u}{s} \tag{1}$$

where z represents the transformed range, x the original range of values, u and s the average and standard deviation of x, respectively.

Another example of pre-processing are transformations based on power laws, responsible for making asymmetric data distributions closer to Gaussian behaviour, thus simplifying the creation of models capable of explaining them and predicting behaviours [12].

The Box-Cox transformation, represented by Eq. (2), is an example of a family of power transformations, being restricted to positive data and characterized by the factor λ [13].

$$y(\lambda) = \begin{cases} \frac{y^{\lambda}-1}{\lambda}, se\lambda \neq 0 \\ logy, se\lambda = 0 \end{cases} \tag{2}$$

Given the exponential trend observed in the data dispersions, a linearization was also performed by applying the natural logarithmic function of the output variable. According to Ostertagova [14], this process makes it possible to describe the behaviour of the number of cycles through a multiple regression, such as the one observed in Eq. (3).

$$ln(y) = \beta_0 + \beta_1 x_1 + \beta_2 x_2 + \beta_3 x_3 + \ldots + \beta_k x_k \tag{3}$$

For the treated application, the use of multiple regression in the polynomial case was considered, as can be seen in Eq. (4).

$$ln(y) = \beta_0 + \beta_1 x_1 + \beta_2 x_2 + \beta_3 x_3 + \beta_4 x_1^2 + \beta_5 x_1 x_2 + \beta_6 x_1 x_3 + \ldots + \beta_{19} x_3^3 \tag{4}$$

where the β parameters are obtained through the method of least squares [15].

The model represented by Eq. (4) was then submitted to the training and validation stage and, subsequently, its response, once an exponential transformation was performed, was evaluated.

2.3 Model Training and Validation

Training and validation are, respectively, the steps in which a learning model is adjusted in order to establish a relationship between inputs and output, and its performance is evaluated. These are, therefore, procedures that directly influence the success or failure of a model [16].

To minimize the chances of failure, Xu and Goodacre [17] mention that an efficient strategy for using the selected dataset is fundamental. Uçar *et al.* [16] report that the most basic one consists of separating the data into training and validation sets and then using them for model construction and performance analysis. However, Queipo *et al.* [18] point out that when using a single pair (training set, validation set), the performance of a learning algorithm become dependent on which points are in the training set and which are in the validation set, so it would be more interesting to explore the existing combinations through more sophisticated strategies.

As an example of these strategies, Kohavi [19] mentions cross-validation, which consists of randomly dividing the dataset into k subsets of approximately the same size. Once a subset is selected, among those possible for validation, the others are used for model training, so that the procedure is repeated k times and, at the end, the performance of the model is taken as the arithmetic average of the individual performances in each combination (training set, validation set).

For the developed model, cross-validation with 10 subsets was used, which is a value recommended in the literature as it balances the need to evaluate different combinations with the computational resources required.

3 Results and Discussion

In order to simplify the presentation and analysis of the results of the developed model, this section was divided into two subsections. The first one presents the proposed model, as well as the values of the correlation coefficient (R^2) and maximum deviation, used as a performance criterion. In the second, the numbers of cycles predicted by the models for the selected aluminium alloy are presented.

3.1 Polynomial Regression

Using the tools presented in the previous section, a polynomial model of degree 3 was developed for predicting the number of cycles supported by a mould until a given preventive maintenance is required. The choice of degree 3 was due not only to the good adherence of the data to the curve, but also to the verification of the poor performance in validation data of polynomials with degree greater than or equal to 4. Through Eq. (5) it is possible to observe the mathematical expression that characterizes the model

$$\begin{aligned} ln(C) \approx 11,990 + 0,535 \cdot D + 0,410 \cdot R + 0,496 \cdot M - 0,08458203 \cdot D^2 \\ -0,004 \cdot D \cdot R + 0,160 \cdot D \cdot M + 0,107 \cdot R^2 - 0,140 \cdot R \cdot M - 0,084 \cdot M^2 \\ +0,003 \cdot D^3 - 0,036 \cdot D^2 \cdot R - 1,162 \cdot D^2 \cdot M - 0,046 \cdot D \cdot R^2 + 2,294 \cdot D \cdot R \cdot M \\ -0.193 D \cdot M^2 - 0,033 \cdot R^3 - 1,180 \cdot R^2 \cdot M + 0,173 \cdot R \cdot M^2 + 0,150 \cdot M^3 \end{aligned} \tag{5}$$

where D, R, M represent, respectively, the variables hardness, average strength and maintenance number after the power transformation and normalization steps, and C the number of cycles until maintenance M is performed. Table 4 shows the average performance of the model in the training and validation steps.

As shown in Table 4, the model represented a good adjustment of the data, with emphasis on the high values of the correlation coefficient and the behaviour of the

Table 4. Result of the prediction model

Training R^2	Maximum Training Deviation	Validation R^2	Minimum Validation Deviation	Maximum Validation Deviation
0,99725 ± 0,00027	14,98%	0,9936 ± 0,0033	0,07%	31,36%

deviation, which in most cases was less than 10% and whose maximum was equal to 31, 36%. This is an indication not only of the adherence of the function to the data, but also the ability to predict unused data for making the curve.

3.2 Estimate for Aluminum Alloy 7034-T6

Aluminum alloy 7034-T6 was chosen to predict the quantity of cycles required for each maintenance, as well as the lifespan of the aluminium mould, since it is a material that, in addition to being recommended for the manufacture of moulds, its hardness and resistance properties are within the range in which the model was validated. Table 5 [20] shows the properties of the mentioned alloy.

Table 5. Aluminum alloy 7034-T6 properties

Average Resistance (kgf/mm^2)	Average Brinell Hardness (HB)
74,4	210

Table 6 shows the number of cycles estimated by the proposed model for that material.

The results provided by the model are consistent when compared to those provided for the steel moulds. Considering its average strength and Brinell hardness values, alloy 7034-T6 was expected to have a lifespan and a quantity of cycles until maintenance within the range determined by steels 1045 Ecofast and Protomold 1050, as can be seen in the graph from Fig. 4.

Since PMo Revolution steel (P20M)(1) has higher resistance and hardness values compared to the mentioned materials, and considering that P20 steel is the most used steel in the manufacture of injection moulds, it is possible to compare the results found for the number of cycles with those estimated for alloy 7034-T6. This is shown in Fig. 5.

It should be noted that, although the 7000 series alloys are the most used, given their good resistance to fatigue and machinability, the moulds manufactured with these materials are less resistant to corrosion compared to the others, making them susceptible to stress corrosion cracking, as discussed by Startks [7]. In this way, it is necessary to carry out treatments and repairs in the moulds, aiming to correct the structural damage that the material may present due to the efforts that it was submitted. It is no coincidence that, when observing the data in Tables 2 and 3, a decreasing variation can be seen between one maintenance and another until the fourth maintenance. This behaviour indicates

Table 6. Number of cycles foreseen by the models for the Aluminum alloy 7034-T6 for maintenance.

Maintenance	Number of Cycles
Maintenance 1	17312,61
Maintenance 2	32871,31
Maintenance 3	49128,42
Maintenance 4	64738,51
Maintenance 5	79272,89
Maintenance 6	93027,98
Maintenance 7	106789,87
Maintenance 8	121692,96
Maintenance 9	139206,73
Maintenance 10	161259,67
Maintenance 11	190525,73
Sum of all maintenance	1055826,68

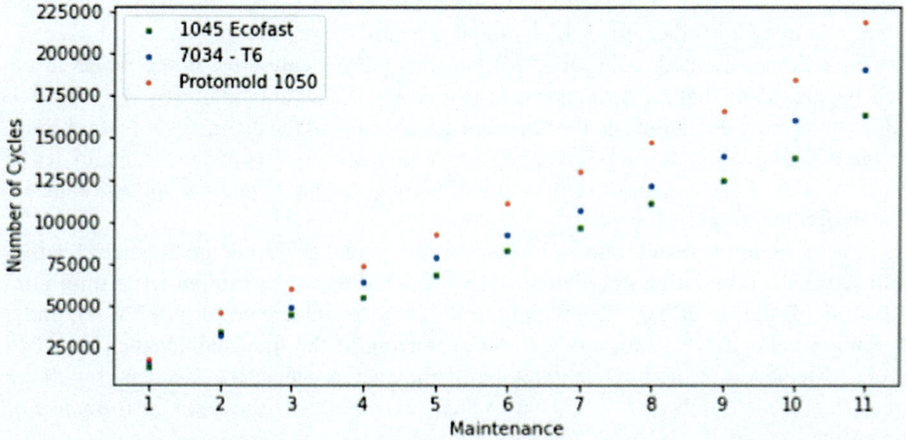

Fig. 4. Estimated number of cycles for each maintenance of alloy 7034-T6

that from the fourth maintenance onwards, treatments or repairs may have been carried out on the steel mould, such as chrome plating, nitriding, weld bead deposition (TIG welding), among other processes [21].

Consequently, the expected result for alloy 7034-T6 also considers the carrying out of treatments in the mould during the exposed period, such as anodizing, chemical nickel bath and low-temperature physical vapor deposition (PVD) [22]. According to Menges and Mohren [1], these treatments not only increase resistance to corrosion, but also resistance to wear and abrasion and, sometimes, facilitate the demoulding of the part.

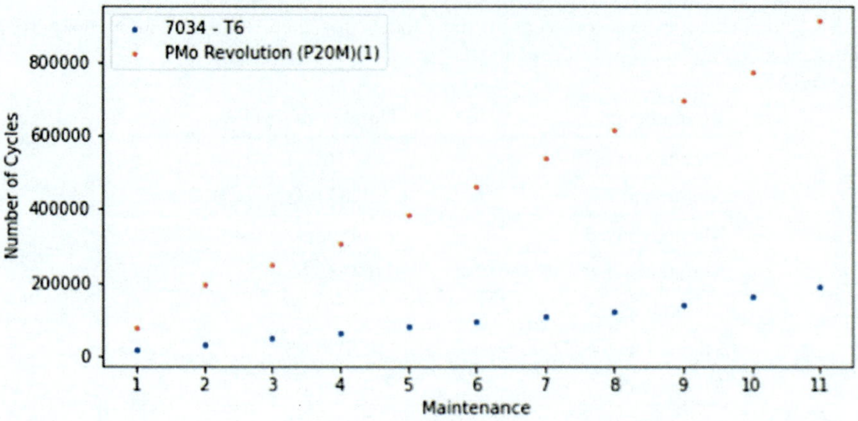

Fig. 5. Estimated number of cycles for alloy 7034-T6 and PMo Revolution steel (P20M)(1)

4 Conclusion

Assessing the maintenance frequency and service life of a mould is an essential step in the performance analysis of injection moulds. In this context, the lack of knowledge regarding the behaviour of aluminium moulds is a barrier to be overcome if the large-scale use of this material is desired in the national scenario.

The model presented in this article is not only a provocation for future works in the area, but also a tool that can be integrated in the study of injection moulds, given the small values of training deviation and the apparent consistency of the estimates for aluminium. After all, the equation developed can provide, under certain circumstances, estimates as to the number of cycles supported by a mould before preventive maintenance is required, and predict the mould's lifespan.

According to the results shown in Table 4, the model presented good generalization ability, with a correlation coefficient value of 99.36%, accompanied by a minimum validation deviation of 0.07%, referring to the sixth maintenance for 1045 steel; and a maximum validation deviation of 31.36%, referring to the first maintenance for 1045 EcoFast steel. Analysing the results globally, the average deviation associated with the prediction of the number of cycles was 4.88%, corroborating the good performance of the proposed model.

The results also revealed that the aluminium 7034 - T6 has a better performance than the steel 1045 Ecofast, since it presented a lifespan 11.59% higher. However, in comparison with PMo Revolution steel (P20M)(1), aluminium performance was 80.77% lower. It is important to point out that, as the productivity expected for the aluminium mould is higher, this information must be considered in this comparative analysis in order to have an estimate of the lifespan in terms of the number of injected parts, thus being more significant information.

References

1. Menges, G., Mohren, T.: How To Make Injection Molding. Hanser Publication, New York (1983)
2. Arieta Filho, F.G. Marchina, P.M. Dihlman, C. Gonçalves, M.: Aplicação de ligas de alumínio em moldes para injeção de termoplásticos. Revista Ferramental (2019)
3. Table of No. of injections, Aço Especial. https://www.acoespecial.com.br/tabela-numero-injecoes.php. Accessed 15 Feb 2022
4. Gardner, E.S., Jr.: Exponential smoothing: the state of the art – part II. Int. J. Forecast. **22**, 637–666 (2006)
5. Penha, O.K.C.: As funções exponenciais e suas aplicações no modelo epidêmico SIR, pp.1–53 (2021)
6. Xu, F., Huang, Z., Yang, F., Wang, D., Tsui, K.L.: Constructing a health indicator for roller bearings by using a stacked auto-encoder with an exponential function to eliminate concussion. Appl. Soft Comput. **89**, 1–15 (2020)
7. Startks, D.: Aluminum Welding. MoldMaking Technology (2016)
8. Williamson, D.F., Parker, R.A., Kendrick, J.S.: The box plot: a simple visual method to interpret data. Ann. Intern. Med. **110**(11), 916–921 (1989)
9. Kotsiantis, S.B., Kanellopoulos, D., Pintelas, P.E.: Data preprocessing for supervised leaning. World Acad. Sci. Eng. Technol., Int. J. Comput. Inf. Eng. **1**(12), 4104–4109 (2007)
10. Mustaffa, Z., Yusof, Y.: A comparison of normalization techniques in predicting dengue outbreak. Int. Conf. Bus. Econ. Res. **1**, 345–349 (2010)
11. Halim, K.N.A., Jaya, A.S.M., Fadzil, A.F.A.: Data pre-processing algorithm for neural network binary classification model in bank tele-marketing. Int. J. Innov. Technol. Explor. Eng. **9**(3), 272–277 (2020)
12. Raymaekers, J., Rousseeuw, P.J.: Transforming variables to central normality. Mach Learn (2021)
13. Li P.: Box-Cox Transformations: An Overview. pp. 1–45. https://www.ime.usp.br/~abe/lista/pdfm9cJKUmFZp.pdf. Accessed 15 Feb 2022
14. Ostertagova, E.: Modelling using polynomial regression. Procedia Eng. **48**, 500–506 (2012)
15. Mahaboob, B., Venkateswarlu, B., Narayana, C., Ravi sankar, J., Balasiddamuni, P.: A treatise on ordinary least squares estimation of parameters of linear model. Int. J. Eng. Technol. **7**, 518–522 (2018)
16. Uçar, M.K., Nour, M., Sindi, H., Polat, K.: The effect of training and testing process on machine learning in biomedical datasets. Math. Probl. Eng. **2020**, 17 (2020)
17. Xu, Y., Goodacre, R.: On splitting training and validation set: a comparative study of cross-validation, bootstrap and systematic sampling for estimating the generalization performance of supervised learning. J. Anal. Test. **2**, 249–262 (2018)
18. Queipo, N.V., Haftka, R.T., Shyy, W., Goel, T., Vaidyanathan, R., Tucker, P.K.: Surrogate based-analysis and optimization. Prog. Aerosp. Sci. **41**, 1–28 (2005)
19. Kohavi, R.: A study of cross-validation and bootstrap for accuracy estimation and model selection. In: Proceedings of the 14th International Joint Conference on Artificial Intelligence, pp. 1137–1143 (1995)
20. MatWeb: Material Property Data. https://www.matweb.com/. Accessed 15 Feb 2022
21. Chen, C., Wang, Y., Ou, H., He, Y., Tang, X.: A review on remanufacture of dies and moulds. J. Clean. Prod. **64**, 13–23 (2014)
22. Kaszynski, J.: Choosing Thermally Conductive Tooling Materials. MoldMaking Technology (2002)

Check for updates

Influence of Different Gear Flank Deviation Tolerance Classes on the Contact Pattern

Caio Felipe Siqueira Gomes(✉), Tiago Cristofer Aguzzoli Colombo, and Ronnie Rodrigo Rego

Instituto Tecnológico de Aeronáutica, Praça Marechal-do-Ar Eduardo Gomes, 50, São José Dos Campos/SP, Brazil
caiocfsg@ita.br

Abstract. Mobility electrification advent has affected the vehicle systems' design requirements, especially for the powertrain components. Among the critical fields affecting the functional performance of future powertrain components is their geometrical accuracy. For gears, the necessity of tighter manufacturing tolerances is related to the much higher rotational speeds involved in the electric motor operation than the internal combustion engine. Although the gear flank tolerance classification establishes the limits of tolerable deviations, there is no treatment regarding how different deviation factors can differently influence the dynamic behavior of gears. Therefore, when standards suggest that high-speed gears require improved tolerance classes, all deviation factors are considered a group. In the case of mobility industries like the automotive, tightening tolerance classes represent a challenge. So, the objective of the present study was the assessment of the influence of different gear deviation factors in tooth contact patterns to identify possible different effects among them. So, tooth contact analyses were performed by computational simulations for a gear sample. The influence of manufacturing profile and helix slope deviations of different tolerance classes in the contact pattern was investigated. The results have demonstrated that a class modification in helix slope deviation has a higher impact on the maximum contact pressure than a class modification in profile slope deviation. When assembly deviations are also considered, the distinct influences are intensified. Identifying the most influential deviation parameters allows the gear manufacturing sector not to have to tighter all tolerances to guarantee an adequate e-mobility gear operation.

Keywords: Gears · Manufacturing Deviations · Contact Stress · Electric Mobility

1 Introduction

The mobility sector is facing a transformation, primer influenced by incorporating the so-called "green technologies" as one of the aspects of the ESG Agenda [1]. The new regulations for the reduction of greenhouse gases emission have evidenced electric and hybrid vehicles as some of the candidates to replace internal combustion-powered ones. This tendency of mobility electrification is highly noticeable when looking at the recent

D. de Oliveira et al. (Eds.): COBEF 2023, LNME, pp. 46–54, 2024.
https://doi.org/10.1007/978-3-031-43555-3_5

European Union (EU) approved bans on the sale of new combustion vehicles after 2035. Over 55% of all new cars in the EU are expected to be fully electrified by 2030, while 40% of new vehicles will include hybrid drive technologies combined with combustion engines [2].

Among all the vehicle systems impacted by the electrification advent, the powertrain is undoubtedly one of those with the most significant trade-offs to be solved, particularly in the case of gears. While internal combustion engines commonly operate at speeds up to 8,000 rpm, the electric motor speeds can easily overcome 15,000 rpm, potentially increasing to 30,000 rpm up to the next decade [3–5]. Although no standards directly address the impact of e-mobility on gear design, an AGMA specification for high-speed helical gears has some suggestions about the effects of the high speeds on gear quality requirements [6].

DIN 3962–1 and AGMA standardize the magnitude of the gear deviations between designers and manufacturers with a system of accuracy [7, 8]. The system has 11 accuracy grades, from 1 to 11, of which grade 1 is the highest degree of accuracy and grade 11 is the lowest, so the lower the class number, the tighter the limits of tolerable gear pitch, profile, and helix deviations. As seen in Table 1, AGMA 6011-J14 establishes flank class 5 as the minimum acceptable for high-speed helical gears [6]. As the current automotive sector hardly works with flank classes better than 7 (sometimes 6, in the case of ground gears), the tighter manufacturing tolerances required for high-speed applications such as e-mobility represents increased manufacturing costs and times [9].

Table 1. Suggested flank tolerance classes for high-speed helical gears [6].

Pitch Line Velocity [m/s]	Flank Tolerance Class
35 - 100	5
100 - 150	4
Over 150	3

High control of the gear flank tolerance classes becomes essential for the electric mobility case since they impact other demands of this application. Hjelm et al. studied the influence of manufacturing error tolerances on gear contact pressure, focusing on the pitch and profile slope errors of ISO 1328 tolerance classes 5–8, typically used in the automotive industry [10, 11]. By simulating the gear meshing and calculating the contact pressure, their results showed that the error class is important and whether the error signal is positive or negative. As the higher torque levels and the reduced number of gears involved in the operation of electric vehicles lead to a higher power density necessity, which directly influences the gears' fatigue life, the actuating contact pressures assessment in the design phase requires even more attention [12].

Hjelm et al. expanded the investigation of Hjelm et al., finding links between static transmission error, contact pressure, and manufacturing error tolerances of gears [10, 13]. Due to the elimination of the internal combustion engine, which is the primary Noise, Vibration, and Harshness (NVH) source in vehicles, the transmission noise is unmasked and becomes more easily perceived in electric vehicles, being amplified by the higher rotational speeds [12]. As the peak-to-peak transmission error (PPTE) is closely related to NVH generation in gearing, the tighter control of the tooth distortions during manufacturing can also contributes to transmission noise reduction.

The referred flank tolerance classification system establishes ranges of tolerable gear pitch, profile, and helix manufacturing deviations for each class number. There is no straightforward treatment regarding how different deviation factors can differently influence the dynamic behavior of gears. If the AGMA 6011-J14 suggestion for the tolerance classes of high-speed gears is rigidly met, all deviation factors tolerances are considered a single group and have to go through tight tolerance control [6]. However, if the deviation types have distinct levels of influence in the efforts involved during gearing, a manufacturing process focused on meeting the tightened tolerance range only for the most influential factors can prevent unnecessary cost increases. Therefore, there is an urgent need to rethink the gear design stage and the manufacturing aspects of the transmission systems themselves.

Given all the above mentioned, the objective of the present study was the assessment of the influence of different gear deviation factors in tooth contact patterns to identify possible different effects among them. Based on this understanding, the concept of design for manufacturing is discussed for the e-mobility gears case.

2 Methodology

2.1 Gear Geometry Selection

Given the strong motivation related to the automotive industry challenges, the gear geometry used for the investigations of the manufacturing deviations impacts in gearing contact pressures was selected to adequately represents this sector. In that way, the *ITA Geometry* was chosen, as its design considers the automotive gears application of both vehicles with internal combustion engines and accompanies the electric vehicle trends in increasing loading conditions.

For the development of the cited geometry, Carvalho mapped the loading operation conditions of the current Brazilian automotive scenario, considering values for maximum power and maximum torque representative of almost 70% percent of all vehicles sold in 2018 [14, 15]. With this procedure, *ITA Geometry*'s nominal power and torque were defined as 110 hp and 240 N·m, respectively [14]. Its main macrogeometry and microgeometry parameters are presented in Table 2 for pinion and gear.

Table 2. *ITA Geometry*'s main parameters [14].

Part	Normal module [mm]	Number of teeth	Normal pressure angle [°]	Helix angle [°]	Reference diameter [mm]	Face width [mm]	Tip relief [um]	Lead crowning [um]	Profile crowning [um]
Pinion	2.85	21	17.5	0	59.85	24	25	6	15
Gear	2.85	28	17.5	0	79.8	27	25	6	15

2.2 Flank Tolerance Classes Selection

Two classes were selected to evaluate the effects of a flank tolerance class variation of distinct gear tooth manufacturing deviations in the flank contact loads. Based on the current automotive sector's regular classes, the first class to be considered was 7. The second class was chosen to represent the e-mobility demands for tighter tolerance control in gears. For a rotational speed of 15,000 rpm, among the operating range for electric motors applications, the pitch line velocity for the *ITA Geometry* is approximately 47 m/s. Accordingly to AGMA 6011-J14, the recommended flank tolerance class is 5 [6]. So, the ISO 1328–1 flank classes 7 and 5 were defined for the analysis [11].

Among the various gear flank deviation factors, profile ($f_{H\alpha}$) and helix slope deviation ($f_{H\beta}$) parameters were chosen to represent, respectively, the involute profile direction and the tooth width direction along the gear flank. By comparing their impact in the contact loads due to a tolerance class variation, it is possible to induce the direction with higher impacts in the stresses involved in tooth contact.

Profile deviations are related to the amount by which an actual profile deviates from the designed one, assessed along with the profile evaluation range (L_α). This range corresponds to the one where the specified accuracy grade is applied. A mean profile of a measured flank, obtained by the least-squares method, aids the evaluation of the deviations in the profile direction [11]. The profile slope deviation ($f_{H\alpha}$) represents the difference between the actual and the designed pressure angle difference. It is verified as the distance between the mean profile and the theoretical involute within L_α, as represented in Fig. 1.

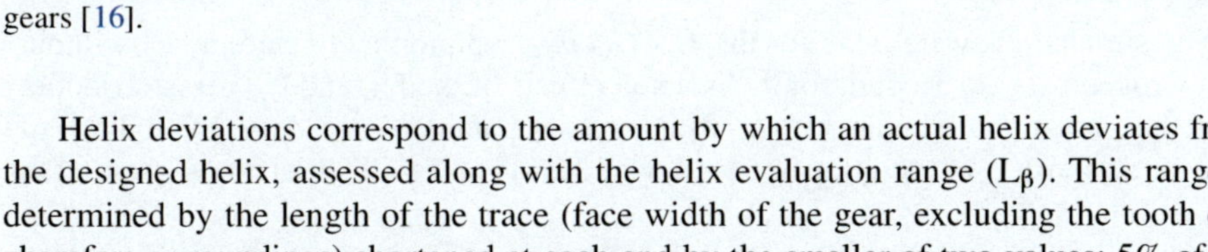

Fig. 1. Definition of the profile slope ($f_{H\alpha}$) and helix slope manufacturing deviations ($f_{H\beta}$) in gears [16].

Helix deviations correspond to the amount by which an actual helix deviates from the designed helix, assessed along with the helix evaluation range (L_β). This range is determined by the length of the trace (face width of the gear, excluding the tooth end chamfers or roundings) shortened at each end by the smaller of two values: 5% of the face width (b) or the length equal to one module [11]. A mean helix of a measured flank, also obtained by the least-squares method, is used to evaluate the distortions in the helix [11]. As shown in Fig. 1, the helix slope deviation ($f_{H\beta}$) represents the actual and designed helix angle difference. It is evaluated as the distance between the mean helix and the theoretical flank line within L_β.

The tolerance limits defined by the AGMA standard for $f_{H\alpha}$ and $f_{H\beta}$ according to the flank class are calculated, in micrometers, by Eqs. (1) and (2) as follows [8].

$$f_{H\alpha} = (0.4m_n + 0.001d + 4)\left(\sqrt{2}\right)^{A-5} \tag{1}$$

$$f_{H\beta} = (0.05\sqrt{d} + 0.35\sqrt{b} + 4)\left(\sqrt{2}\right)^{A-5} \tag{2}$$

In the above expressions, m_n represents the normal module of the gear in millimeters, d is the reference diameter in millimeters, b is the gear face width in millimeters, and A is the flank tolerance class number. Rounded rules for these parameters defined by AGMA state that [8]:

- If the value is greater than 10 μm, the round must be to the nearest integer micrometer;
- If the value is 5.0 μm or greater but less than or equal to 10 μm, the round must be to the nearest 0.5 μm;
- If the value is less than 5.0 μm, the round must be to the nearest 0.1 μm.

Besides comparing the $f_{H\alpha}$ and $f_{H\beta}$ class alteration impacts in the contact between teeth, additional analysis of the flank contact stress patterns was done by adding assembly deviations of the geared pair. Both in-plane deviations ($f_{\Sigma\delta}$) and out-of-plane deviations ($f_{\Sigma\beta}$), schematized in Fig. 2, were considered to identify possible combined effects with the flank deviations on the actuating stress heterogeneities along the gear flank.

2.3 Tooth Contact Analysis

The assessment of the impacts of varying the flank class number for the selected manufacturing deviation types was performed by evaluating the *Hertzian* pressures and the flank contact patterns resulting from each condition. For this, tooth contact analysis of the *ITA Geometry* pair was done on the *KISSsoft* software. This software gathers several standards in a unique interface and allows the execution of numerous computational simulations involving various factors combinations, being commonly used for machine elements design.

The simulations were done for the *ITA Geometry* pinion and gear contact without the microgeometry to account for the isolated effects of the $f_{H\alpha}$ and $f_{H\beta}$ class variation. After that, additional simulations of flank contact patterns were done by adding in-plane deviations ($f_{\Sigma\delta}$) and out-of-plane deviations ($f_{\Sigma\beta}$). In all simulations, the torque

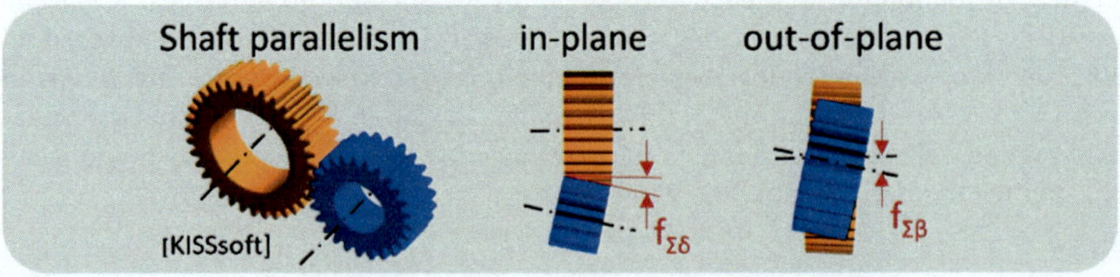

Fig. 2. Types of assembling deviations of a geared pair.

level considered was *ITA Geometry*'s nominal value of 240 N·m at a rotational speed of 15,000 rpm.

3 Results and Discussions

Through Eqs. (1) and (2), the tolerance ranges for $f_{H\alpha}$ and $f_{H\beta}$ for classes 7 and 5 were calculated for each gear. For class 7, the determined ranges were, respectively, $\pm$ 10 μm and $\pm$ 12 μm for the pinion's $f_{H\alpha}$ and $f_{H\beta}$, and $\pm$ 10 μm and $\pm$ 13 μm for the gear's $f_{H\alpha}$ and $f_{H\beta}$. In turn, the class 5 range for $f_{H\alpha}$ and $f_{H\beta}$ were calculated, respectively, as $\pm$ 5 μm and $\pm$ 6 μm for the pinion and as $\pm$ 5 μm and $\pm$ 6.5 μm for the gear.

Figure 3 shows the results obtained by the tooth contact computational simulations varying the flank tolerance class of the manufacturing deviations, where index 1 is related to the pinion and index 2 to the gear. The map evidences the impact of a combination of values of $f_{H\alpha}$ and $f_{H\beta}$ in the maximum *Hertzian* contact pressure of the gears. The global axes are related to the pinion deviations, while the 25 internal squares constituting the global map are related to the gear deviations. By analyzing the profile and helix slope class variation effects, it can be seen that, for both *ITA Geometry*'s pinion and gear, the latter deviation parameter has higher effects on the contact pressure than the further. It is concluded that a flank tolerance class variation from 7 to 5 in $f_{H\beta}$ produces more significative effects in the *ITA Geometry* operating contact stresses than the same variation in $f_{H\alpha}$.

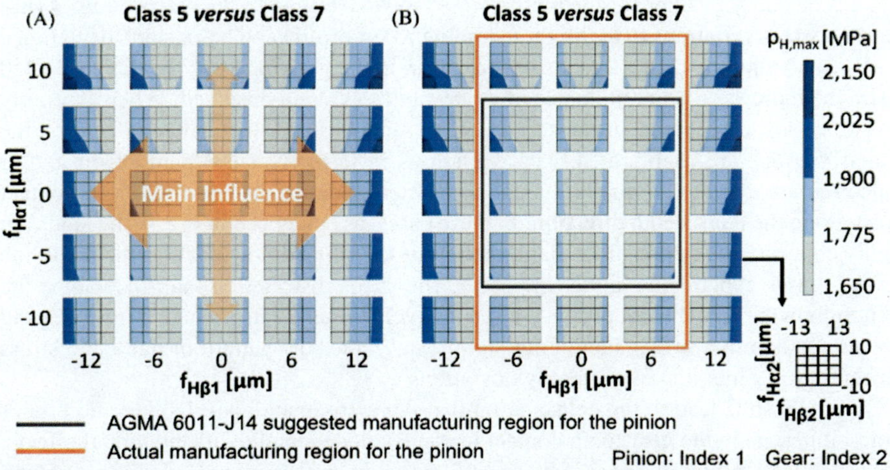

Fig. 3. Effects of class variation of profile and helix slope deviations in the maximum *Hertzian* contact pressure of the *ITA Geometry*. (A) Main influence due to $f_{H\beta}$. (B) Actual and AGMA suggested manufacturing working regions for deviations of the pinion [6].

The region delimited in grey in Fig. 3b indicates the AGMA flank tolerance class 5 recommendation for the pinion at a rotational speed of 15,000 rpm [6]. However, a close look at the map allows the constatation that other combinations of parameters outside the class 5 limits have similar maximum *Hertzian* pressures. Once $f_{H\alpha}$ and $f_{H\beta}$ do not have the same effect on the gear contact pressures, the manufacturer's working region can be expanded to the area delimited in orange, less tight than suggested by AGMA [6]. So, the results evidence that manufacturing might not necessarily need to meet the flank tolerance class recommendation for all factors to meet an adequate operating region.

These findings regarding the inequality effects of the deviation parameters are similar to that observed in the literature. The results of Hjelm et al. and Hjelm et al. evidenced that a gear flank class variation of profile slope error and pitch error do not have the same impacts [10, 13]. Although this constatation was not presented in their studies, an analysis of their results allows the conclusion that, for the FZG C spur gear geometry, a class variation of pitch error has a more significant influence in the maximum gear contact pressure and PPTE than profile slope. As contact pressure and peak-to-peak transmission error are related, respectively, to power density and NVH generation, two of the electromobility concerns for gears, knowing the flank deviations that mainly influence them is an important step in a design for e-mobility.

Although the analysis of tolerance class variation impacts on the intensity of the tooth contact pressures allowed important conclusions, it has to be complemented by an assessment of how these actuating stress are distributed along the entire tooth flank. Stress heterogeneities along the flank may indicate susceptible regions to failure initiation. To evaluate how profile and helix slope deviations influence the stress distribution in the flank, three combinations of deviation values were chosen among those presented in Fig. 3. For all of them, $f_{H\alpha 1}$ and $f_{H\beta 1}$ were fixed, respectively, in 10 μm and -12 μm. Then, three cases were selected as different combinations of $f_{H\alpha 2}$ and $f_{H\beta 2}$, as shown in Fig. 4. The chosen values for the pinion aimed to get a region of the map on Fig. 3 with elevated stress variations to highlight the influence of profile and helix slope deviations.

By analyzing the contact pattern distribution in the gear flank for the three selected cases, the more significant influence of $f_{H\beta}$ variation is again observed. When assembly deviations are considered in the tooth contact analysis, the inequality effects of the gear deviations are highlighted. The deviations in terms of shaft parallelism and/or deflection are expected to naturally lead to a heterogeneous distribution of the contact stress along the flank width direction. In these cases, as evidenced in Fig. 4, the loads are highly concentrated in one of the flank extremities. As $f_{H\beta}$ also actuates in the flank line direction, this explains its higher impact on the stress distribution than $f_{H\alpha}$. Since the points along the teeth height profile are successively brought into contact during gearing action, an alteration in $f_{H\alpha}$ should not significantly alter the pattern of flank and stress concentration generated by assembly deviations.

All the results lead to the constatation that there are manufacturing deviation types more influential in the gear tooth contact loads. By understanding its individual effects, it is possible to rethink the gear tooth tolerance design and the manufacturing process to solve the trade-offs brought by the electric mobility application regarding the necessity of gears with higher qualities.

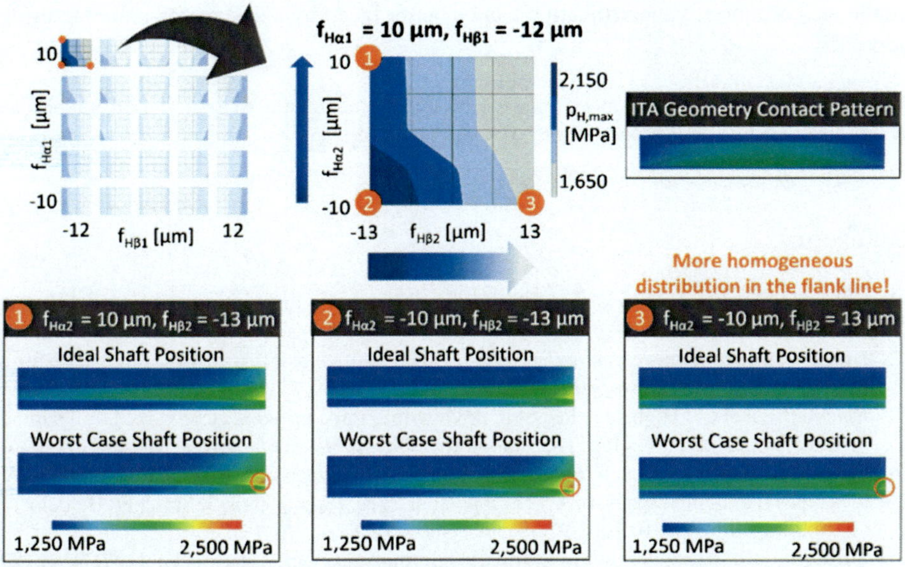

Fig. 4. Influence of profile and helix slope deviations in the contact pattern of the *ITA Geometry* gear pair considering the existence of assembly deviations.

4 Conclusions

The study evidenced that the distinct gear tooth manufacturing deviation types have different impacts on the maximum contact stresses and their distribution along the gear flank. Specifically, an alteration of gear flank tolerance class for the helix slope deviation was demonstrated to have a higher impact on the stress intensity and distribution than the same class alteration in profile slope deviation for the *ITA Geometry*. Although the results were generated for the profile slope and helix slope parameters, which are related, respectively, to distortions in the involute tooth profile and gear flank width directions, it is possible to qualitatively extrapolate them to other parameters that also correspond to these directions. In that sense, it is concluded that the manufacturing errors in the flank width direction are more harmful than the ones in the tooth profile direction. When assembly deviations were added to the tooth contact analysis, the distinct behavior of the manufacturing errors was highlighted.

The conclusions of this work are of great interest to the gear manufacturing chain, especially for the ones engaged with the electromobility demands in terms of high accuracy and precision gears. Once the manufacturing distortions have distinct impacts on the gearing contact, a design for tighter tolerances on the most impacting parameters and a manufacturing chain oriented to make it possible need to be discussed. If the attention is directed to reducing the manufacturing errors highly affecting the gear operation behavior, the cost increases in the manufacturing due to tighter tolerances will be lower than the ones in case all deviation limits were reduced as recommended by the AGMA specification for high-speed gears. If designers and manufacturers act together, the powertrain challenges of mobility electrification can be better faced by a design-for-manufacturing approach.

Acknowledgments. The authors gratefully acknowledge the funding support of the Coordenação de Aperfeiçoamento de Pessoal de Nível Superior (CAPES).

References

1. ISS Homepage. https://www.issgovernance.com/library/the-future-of-mobility/. Accessed 05 May 2023
2. PWC. https://www.pwc.com/gx/en/industries/automotive/assets/pwc-five-trends-transforming-the-automotive-industry.pdf. Accessed 05 May 2023
3. Reichert, U., et al.: High speed electric drive with a three-speed gear box. In: Car Training Institute, pp. 25–27. CTI Mag: The Automotive TM, HEV and EV Drives, Dusseldorf (2016)
4. Mileti, M., Strobl, P., Pflaum, H., Stahl, K.: Design of a hyper-high-speed powertrain for EV to achieve maximum ranges. In: CTI Symposium 2018. P, pp. 265–273. Springer, Heidelberg (2020). https://doi.org/10.1007/978-3-662-58866-6_21
5. Stadtfeld, H.J.: Introduction to electric vehicle transmissions. Gear Technol. **37**(7), 42–50 (2020)
6. ANSI/AGMA 6011-J14: Specification for High Speed Helical Gear Units. Alexandria (2014)
7. DIN 3962–1: Accuracy of cylindrical gears: tolerances for individual errors. Berlin (1978)
8. ANSI/AGMA ISO 1328-1-B14: Cylindrical Gears – ISO System of Flank Tolerance Classification – Part 1: Definitions and Allowable Values of Deviations Relevant to Flanks of Gear Teeth. Alexandria (2014)
9. Dizdar, S., et al.: Process, quality and properties of high-density P/M gears. Adv. Powder Metall. Part. Mater. **9**, 9–36 (2003)
10. Hjelm, R., et al.: Influence of manufacturing error tolerances on contact pressure in gears. Proc. Inst. Mech. Eng. C J. Mech. Eng. Sci. **235**(20), 5173–5185 (2021)
11. ISO 1328–1: Cylindrical gears - ISO system of accuracy – Part 1: Definitions and allowable values of deviations relevant to corresponding flanks of gear teeth. Geneva (1995)
12. AGMA White Paper: A Gearing-Centric Snapshot of the EV Space. Alexandria (2021)
13. Hjelm, R., et al.: Gear tolerancing for simultaneous optimization of transmission error and contact pressure. Results Eng. **9**, 100195 (2021)
14. Carvalho, A.A.: Surface integrity evolution of Nb-Ti microalloyed steels along the gear manufacturing chain. 113f, Dissertation of Master of Science in Materials, Manufacturing and Automation – Instituto Tecnológico de Aeronáutica, São José dos Campos (2020)
15. FENABRAVE: Informativos – Emplacamentos, p. 46. São Paulo (2019)
16. Klocke, F., Brecher, C.: Zahnrad- und Getriebetechnik. Hanser, München (2017)

Thermomechanical Evaluation of Thermoplastics Injection Cycle Effects in Aluminium Moulds Using the Finite Element Method

Gimaézio Gomes Carvalho[1](✉), Gabriel Vasconcellos Bayma[2], Márcio André Fernandes Martins[1], Cristiano Vasconcellos Ferreira[3], Valter Estevão Beal[4], and Armando Sá Ribeiro Junior[1]

[1] UFBA, R. Prof. Aristídes Novis, 2 - Federação, Salvador, BA 40210-630, Brazil
gimaezio@hotmail.com, {marciomartins,asrj}@ufba.br

[2] IEL-Ford Motor Company, Edístio Pondé, 342 - Stiep, BA 41770-395 Salvador, Brazil
gvbayma@uol.com.br

[3] UFSC, Campus Universitário, s/n - Trindade, SC 88010-970 Florianópolis, Brazil
cristiano.v.ferreira@ufsc.br

[4] SENAI CIMATEC, Avenue Orlando Gomes, 1845 - Piatã, BA 41650-010 Salvador, Brazil
valtereb@fieb.org.br

Abstract. The mould stiffness becomes more relevant during the packing stage, therefore predicting the mould mechanical behaviour when submitted to the loading cycles of the injection process is essential to the tooling project. Some loading sources are more likely to damage the tooling, such as the holding pressure, the clamping force, and the thermal stress. Considering the mould thermal cycle is important, evaluating heat exchange by conduction between the part and the mould, the free convection with the surrounding air and the forced convection with the cooling fluid. The aluminium's thermal and mechanical properties are significantly different than steel ones, which change the injection cycle and its capability to resist the injection load cycles. This justifies the need to evaluate the differences in the use of both steel and aluminium materials in the mould design. This paper proposes numerical simulations making use of finite element method to assess the effects of the injection moulding loads on aluminium moulds, through outputs such as temperature profile, the stress state, and parting line openings. The results are compared to numerical results obtained for steel moulds. The results show that the difference between the mould's materials properties significantly changes the injection moulding conditions and the mould design, particularly the cooling system efficiency. These aspects are fundamental to make guidance to required modifications on the aluminium moulds, allowing the mould mechanical project and cooling system optimization to perform the best in terms of the part quality and process productivity.

Keywords: Aluminium Moulds · Thermomechanical Analysis · Finite Element Method · Injection Mould

D. de Oliveira et al. (Eds.): COBEF 2023, LNME, pp. 55–69, 2024.
https://doi.org/10.1007/978-3-031-43555-3_6

1 Introduction

It is remarkable the number of industrial sectors that use thermoplastics as raw material for making geometrically complex products, from the injection moulding process [1]. Therefore, it is necessary to constantly produce technical data about this process, highlighting in this work those referring to the mould's resilience and performance.

Considering this scenario, it is also worth stating that for high-volume production, thermoplastic injection moulds are traditionally made of steel due to its good thermal stability and high mechanical strength. Nonetheless, despite these advantages, steel has a low thermal conductivity when compared to other metals, such as aluminium, which results in long injection cycles, mainly due to the longer time required to cool the part before demoulding.

Arieta Filho [2] presents thermal and mechanical properties of four materials that are commonly used in thermoplastic injection moulds. Based on those data, it is possible to affirm that aluminium has a thermal conductivity about four times greater than steel. However, what reduces the use of this material in the mould manufacturing is its low mechanical strength, when compared to steel, causing suspicion about its lifespan, which is measured by the number of parts that can be injected using these moulds.

These mechanical properties discouraged the use of aluminium moulds, which resulted in little information available in the literature on the various aspects related to the use of this type of tool. Nevertheless, the thermal properties suggest a significant gain in the injection cycle, and the aluminium alloys of the 7XXX series have the sufficient mechanical strength to withstand the injection pressures of thermoplastics [2].

Heat transfer within the mould is a complex transient three-dimensional problem. When starting the production, about 10 to 20 cycles are necessary to establish a steady-state regime, at which point the mould temperature evolution is the same for successive cycles [3]. During the injection moulding process, the moulds are submitted to three major loads: thermal stress due to heat transfer; stresses due to injection pressure imbalance; and the clamping force induced stress. In general, the thermoplastic injection moulding cycle consists of the following stages: mould closing; filling; packing; cooling; mould opening; and part ejection. Some of these steps are not relevant to be evaluated in terms of ensuring the mould design withstands the clamping force and the cavity pressure, for example, mould closing/opening and part ejection.

Chen *et al.* [4] investigated the influence of processing parameters (including injection speed, melt temperature, mould temperature, filling–packing switchover and packing pressure) on the mould separation under different clamping forces. The associated variations of part weight were also measured. A simplified three-dimensional finite element model of mould structure on the core side plate was used for structural analysis. As there are interfaces, either between the mould and the part or between the mould itself, the thermal conductance becomes important in a simulation to better describe the heat transfer through these regions [5].

On the other hand, monitoring this property involves a complex and error-sensitive experimental procedure [3]. Somé *et al.* [6] showed that the thermal conductance relies on the roughness of the mould and the polymer. Experiments revealed different thermal conductance values for a polypropylene-steel interface, considering several surface roughness and pressures. These data are applied in this work.

The main difference between this study and those found in the literature is the thermomechanical evaluation of using aluminium in complex injection moulds. This paper considers a full three-dimensional mould model, in order to assess items like temperature profile, stress state and parting line openings. The authors are unaware of a work that modelled all these details through the finite element method, e.g., main components are fully modelled, forced convection in the cooling channels, natural convection with the surrounding air, thermal conductance in the interfaces.

2 Methodology

This paper proposes to simulate successive end-to-end injection cycles until steady state has been reached, then evaluate the stress state and parting line openings. This procedure is necessary to ensure a more representative temperature profile and, consequently, an accurate thermal stress state. The total cooling time was set as a variable through transient numerical simulations to determine the influence of this parameter on mould stress state.

This methodology was established to aim at thermomechanical evaluations, through numerical simulations, regarding effects in injection moulds arising from processing parameters variation. Particularly, the case study is directed to using aluminium as the mould material, when used for high-volume production. Evaluations changing only the mould material to steel are made to obtain a direct comparison between the two materials.

2.1 Geometry

For this work, a four parts mould (cavity plate, cavity, core plate and core) plus the moulded part were considered. The mould was designed to have four cooling channels on the cavity side (cavity + cavity plate) and two cooling channels on the core side (core + core plate), being aligned through four guide pins. There are also two holes that represent temperature monitored locations. The mould dimensions are 190.2 mm/246.0 mm/346.0 mm. The moulded part thickness is around 2.0 mm and represents a vehicular cupholder. Figure 1 shows the components.

2.2 Numerical Model

To evaluate the forces acting on the mould during the first injection cycle, the initial step is the transient heat transfer simulation up to end of the packing stage. At the beginning, uniform initial temperatures are established, both for the moulded part and the mould, with the thermomechanical simulation starting from a free strain state. The effect of plastic flow into the cavity during the filling stage is disregarded. The intermediate step of the simulation is a simultaneous application of clamping force and holding pressure, considering thermal strains from initial step. This step is performed through a static simulation. At the end of the intermediate step, both mould's stress state and the parting line openings are assessed. The final step consists of simulating the heat transfer until the mould opening. Boundary conditions must be properly imposed to faithfully represent what happens in the injection moulding machine. Gravity must be applied.

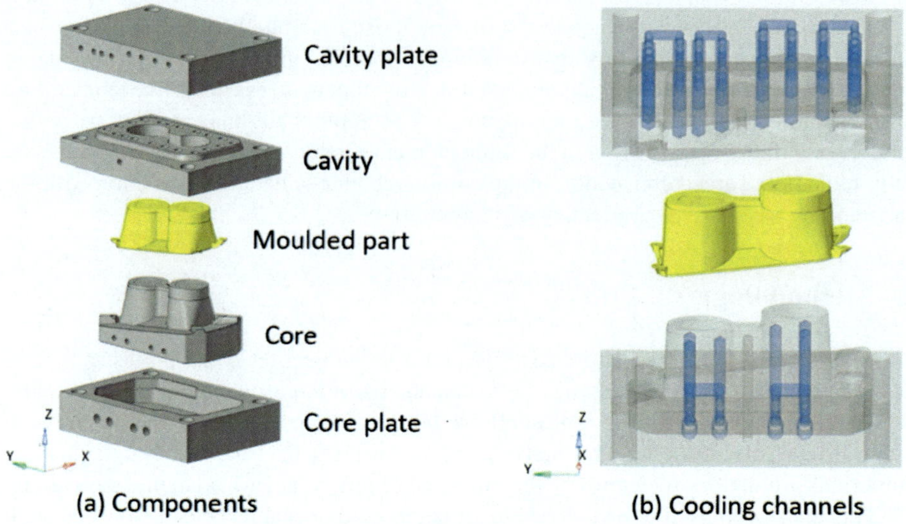

Fig. 1. Mould and moulded part geometries.

Heat transfer by conduction is the predominant phenomenon during the cooling, thus contact surfaces are used at the model interfaces to represent the solid-solid interaction, for both the thermal and mechanical purposes. Heat transfer by convection occurs simultaneously on the mould external surfaces – with the air – as well as in the cooling channels, so surfaces are defined for the adequate representation of this phenomenon. For the mechanical interaction, the friction coefficient is set, both for the polymer-mould interface and the mould's components interfaces. For the thermal interaction, suitable thermal conductance values are applied.

To simulate the second injection cycle, the same procedure is adopted, but assuming as an initial condition for the mould the final temperature profile of the previous cycle. For the moulded part, the application of a uniform initial temperature is kept, representing the end of the filling. This procedure is repeated for successive cycles until a stopping criterion that represents a steady state is met.

Upon completion of the planned numerical simulations, the temperature profile, stress state, and the parting line openings are verified to evaluate possible tooling damages or moulded part issues. This approach is schematically presented in Fig. 2.

A solver based on Finite Element Method was used to run the coupled temperature-displacement, static, and uncoupled heat transfer simulations. It was built a mesh with 5684930 tetrahedral elements and 1107297 nodes. Table 1 shows the material properties that were applied in this work, both to the moulds and to the moulded part. These mechanical properties were obtained from the data sheet of the material suppliers and are related to Steel P20 and Aluminium 7075.

In relation to the properties applied at the interfaces, the data were estimated by Somé *et al.* [6]. To estimate the film coefficient with the surrounding air, correlations for natural convection in external flows were used. To estimate the film coefficient with the cooling fluid (water), correlations for forced convection in internal flows were used.

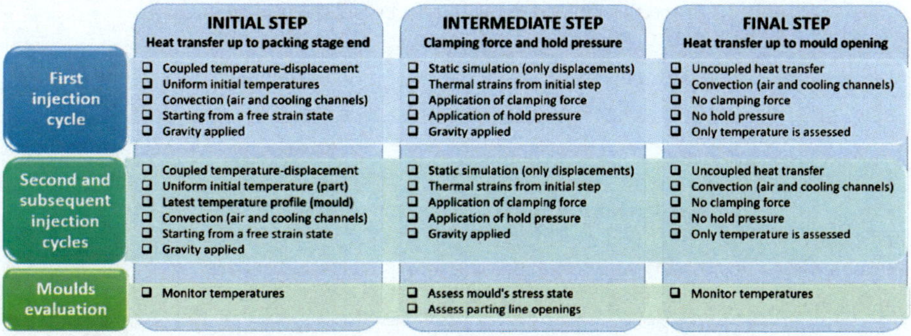

Fig. 2. Approach to evaluate injection moulds through thermomechanical numerical simulations.

Table 1. Steel, aluminium, and polypropylene properties.

	Steel	Aluminium	Polypropylene
Density (kg/m^3)	7890	2800	883
Poisson's ratio	0.30	0.34	0.38
Coefficient of thermal expansion (°C^{-1})	0.000011	0.0000235	0.00015
Young's modulus (MPa)	210000	82116.6 at 82.22 °C 79664.0 at 117.8 °C 77609.0 at 160.9 °C 75689.3 at 212.5 °C	1700.0 at −40.0 °C 1000.0 at 23.0 °C 300.0 at 80.0 °C 100.0 at 220.0 °C
Specific heat [J/(kg · °C)]	452	931.45	2800
Thermal conductivity [W/(m · °C)]	48	136.87	0.15

For this last one, a Moldex3D® preliminary simulation was necessary to estimate the Reynolds number. Table 2 shows the properties applied at the interfaces as well as the film coefficients.

2.3 Evaluation of the Processing Parameters Contributions

As Huang *et al.* [7] described, the clamping force can be estimated by multiplying the projected mould cavity area and the injection pressure. To verify the influence of these parameters in the tooling stress state and parting line separation, a set of processing parameters derived from Moldex3D® successful rheological simulations was used, as shown in Table 3.

The pressure profile that was applied to cavity elements was also obtained from Moldex3D® rheological simulation. The pressure ranges from 30 MPa (gate) to 24.3 MPa (point away from the gate), as shown in Fig. 3.

Table 2. Properties applied to the interfaces and the film coefficients.

	Polymer-mould interface	Mould-mould interface	Surrounding air	Cooling fluid (water)
Friction coefficient	0.4	0.1	-	-
Thermal conductance [W/(m2 · ºC)]	4000.0 (0.0 mm) 1500.0 (1.0 mm) 0.0 (>1.0 mm)	4000.0 (0.0 mm) 1500.0 (1.0 mm) 0.0 (>1.0 mm)	-	-
Film coefficient [W/(m2 · ºC)]	-	-	6.5	11433

Table 3. Processing parameters used initially in the simulations.

Melt temperature	Initial mould temperature	Holding time	Holding pressure	Clamping force	Cooling time	Total time
220 ºC	20 ºC	6 s to 10 s	From rheological simulations	650 kN	12 s to 16 s	18 s to 26 s

Fig. 3. Maximum pressures verified from Moldex3D® rheological simulation.

Hereby, it was set an evaluation plan to better understand how significant the cooling is on the temperature profile, stress state and mould parting line opening. The cooling time and the holding time are permuted using 12 s, 14 s and 16 s for cooling plus 6 s, 8 s and 10 s for holding. Therefore, the total time is the sum of holding time and cooling time. The filling stage was disregarded in this work.

3 Results

This section contains the results to aluminium and steel mould simulation, such as the temperature profile, stress state, and parting line openings.

3.1 Temperature Profile

Three temperature histories were recorded, named locations A (core side) and B (cavity side), in addition to the average temperature of the moulded part, as shown in Fig. 4. The average temperature is the sum of all temperatures divided by the number of part nodes. The temperature results for both moulds are presented as a function of time and different cooling conditions.

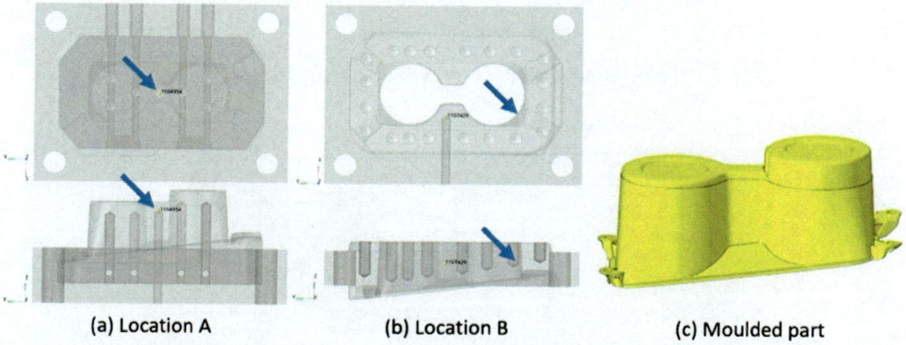

Fig. 4. Node positions where temperature histories were recorded.

Figure 5 shows temperature response comparisons between steel and aluminium moulds for a holding time of 6 s and cooling time of 12 s, totalling 18 s per cycle. Based on the results of the mould internal points and the part average temperature, it can be stated that the cyclic steady state in the steel mould occurs with a greater number of cycles (seven cycles) compared to the aluminium mould (four cycles), reaching higher temperature values but with a smaller thermal amplitude during cycles. When analysing the other cases, this pattern was also observed.

Additionally, analysing the moulded part hottest node during the 10th cycle for a total time of 18 s per cycle, it was observed that for both materials the hotspot final temperatures are very similar, as it was also verified in the rheological simulations. In the case of Polypropylene components, the hottest regions of the product reach about 100 °C when the part is ejected, as mentioned by Ahuett-Garza *et al.* [8]. Therefore, the ejection temperature threshold was stablished at 100 °C. Figure 6 shows the hottest node temperature responses during the 10th cycle for both materials. It is noticed that for the shortest total time (18 s), the moulded part can be ejected from both moulds, being approximately 16 s for aluminium mould and 17 s for steel mould. The reason why this point is almost equal in both moulds is because the hot spot is a region that cools considerably slower than others, due to its heat concentration.

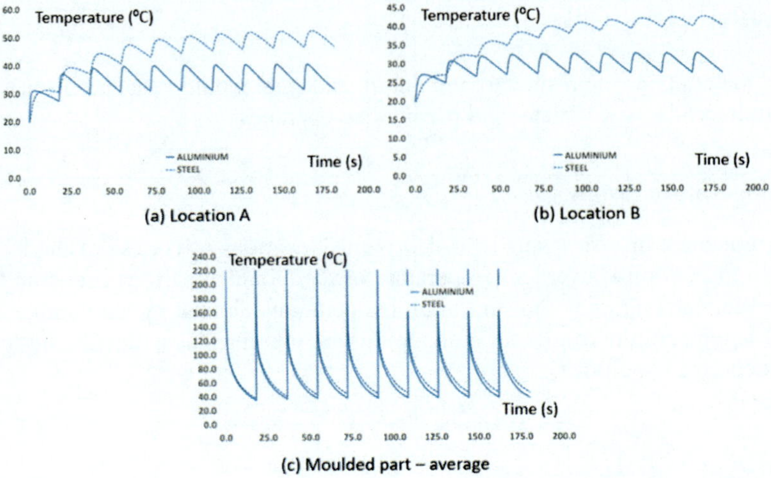

Fig. 5. Temperature responses for a total time of 18 s per cycle.

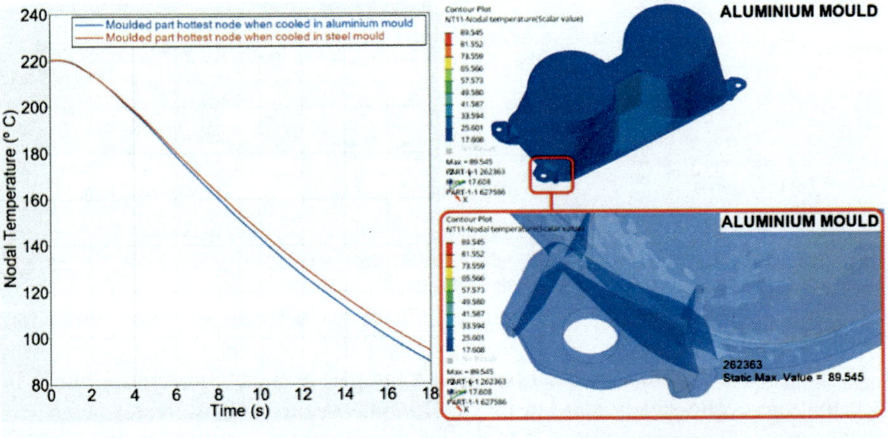

Fig. 6. Moulded part hottest node during the 10th cycle for a total time of 18 s per cycle.

Figure 7 shows the maximum temperature reached during 10th cycle at points A and B as a function of total time. As would be expected, for longer injection cycles the maximum temperature is lower than for faster injection cycles, with steel reaching higher temperatures than aluminium. The longer the injection cycle is, the closer the moulds' temperatures get.

Figure 8 shows the temperatures at the ejection time of the 10th cycle. Based on these data, the moulded part average temperature when cooled in the aluminium mould for a period of 18 s is the same as the part cooled in a steel mould for 24 s, which proves the thermal efficiency of aluminium compared to steel tooling.

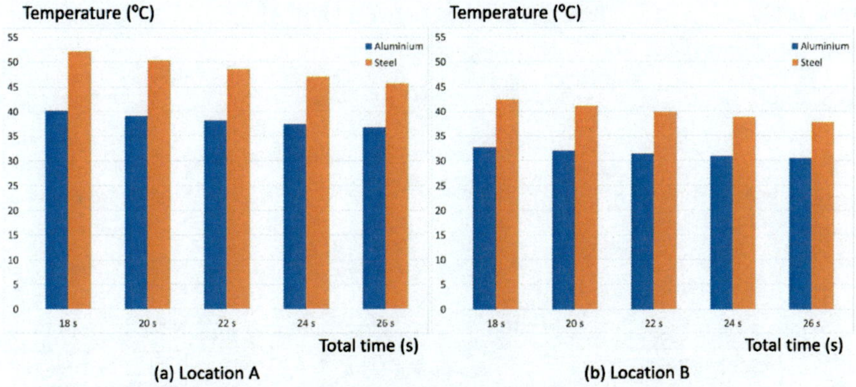

Fig. 7. Maximum temperature responses as a function of total time.

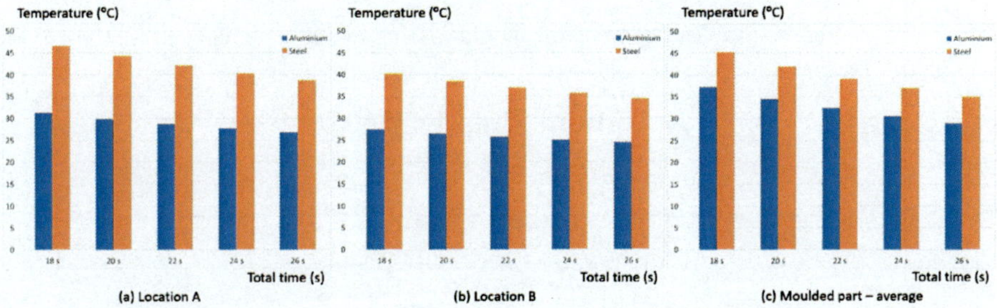

Fig. 8. Temperatures responses at the ejection time as a function of total time.

3.2 Stress State

Figure 9 presents the stress numerical results for both moulds. These data show the maximum principal and shear stresses acting on the mould during the packing stage at the cyclic steady state, which is due to temperature gradient, clamping force and holding pressure. The stress states observed are slightly greater for aluminium mould than for steel mould and becomes greater with the holding time increasing and lesser with the cooling time increasing.

Additionally, to understand how the stress state behaves during the transient regime, maximum stresses were collected for a holding time of 8 s and cooling time of 14 s, totalling 22 s per cycle, as shown in Fig. 10. The observed stresses are greater for the steady state than for the first cycle. The stresses are greater for aluminium mould than for steel mould.

As shown by Arieta Filho [2], P20 steel usually has a yield strength of around 800 MPa. The failure criterion adopted for steel was von Mises (Distortion Energy Theory). The maximum von Mises stress monitored in steel mould was 531.9 MPa, below the yield strength. Aluminium alloy 7075 has a yield strength of about 500 MPa and the failure criterion adopted was Tresca (Maximum Shear Stress Theory). Thus, the target to be compared to the Maximum Shear Stress found (293.8 MPa) in aluminium mould

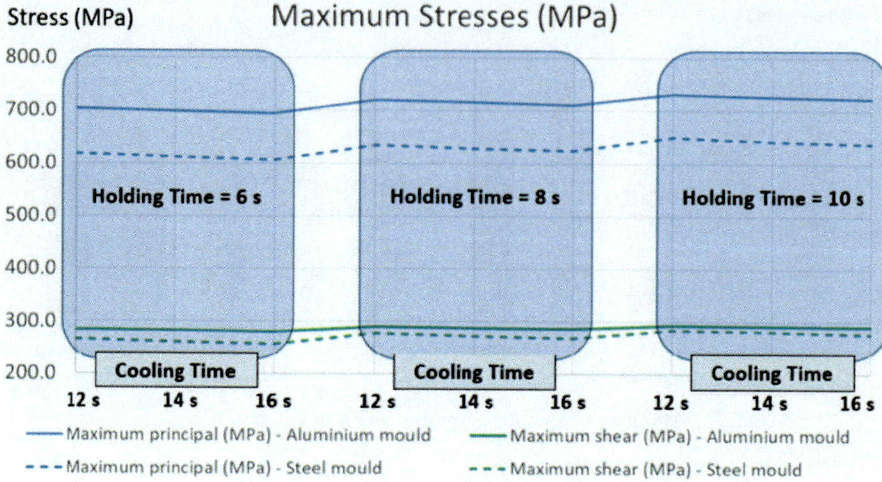

Fig. 9. Stresses acting on the moulds during steady-state regime.

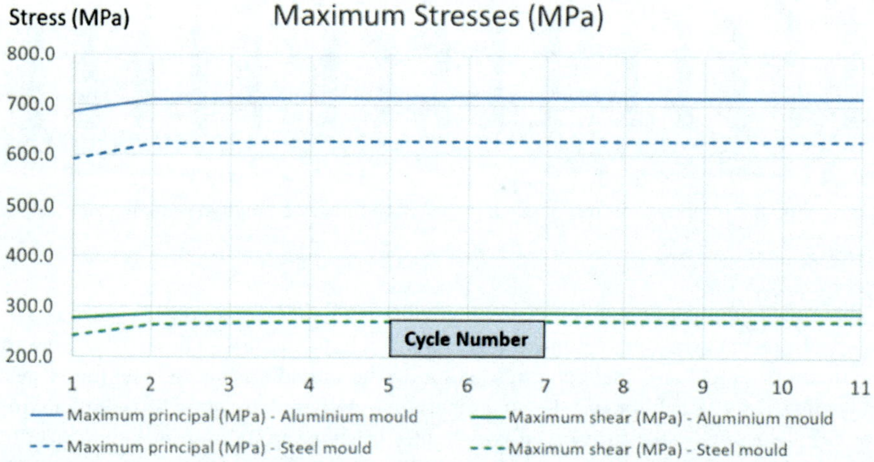

Fig. 10. Stresses acting on the moulds during transient regime for a holding time of 8 s and cooling time of 14 s.

is 250 MPa. Based on these results, the aluminium mould is susceptible to permanent deformations and the steel one is not, and changes in processing parameters or in the mould design are recommended so that aluminium benefits from its thermal advantages without harming the thermoplastic part manufacturing. Figure 11 shows the two elements where the maximum stresses were recorded. They belong to the component Cavity, as defined in Fig. 1.

Three simulations were devised to better understand the influence of each loading conditions: clamping force (C), holding pressure (P), and thermal loading (T). Initially,

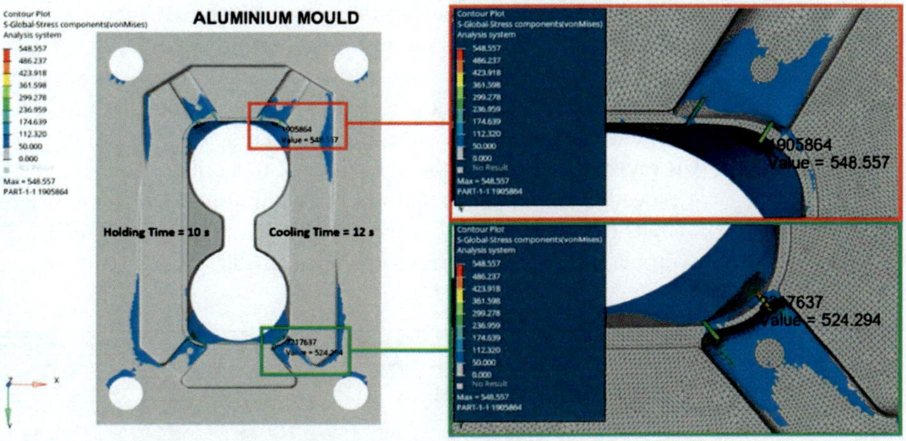

Fig. 11. Mould area where the maximum stresses were recorded.

a static simulation was performed, considering only the clamping force of 650 kN. Subsequently, a new simulation was run considering the clamping force of 650 kN plus the holding pressure (C+P). Finally, the whole coupled temperature-displacement analysis was performed (C+P+T) during the cyclic steady-state regime. The holding time of 8 s and cooling time of 14 s were considered for this detailed investigation. As may be noticed from Fig. 12, the loading condition that increases the stresses of the two critical elements (ID 1905864 and ID 2217637), significantly, was the holding pressure.

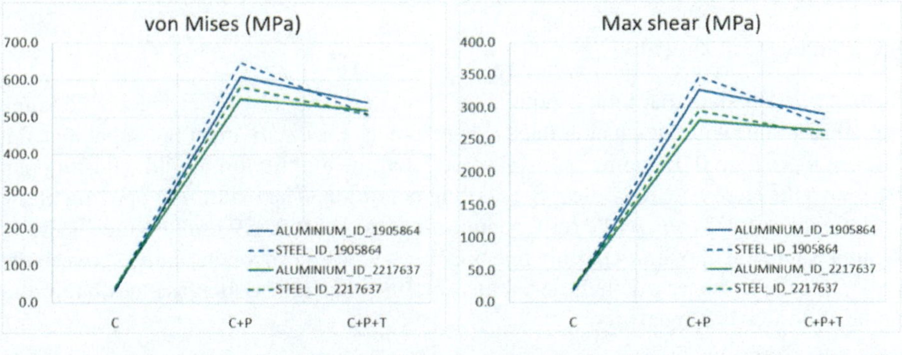

Fig. 12. Loading condition breakdown and stresses evaluation for a holding time of 8 s and cooling time of 14 s.

To evaluate in detail the loading condition contributions, the normal and shear stresses of the worst element (ID 1905864) were collected when cooled into both moulds. Then it was calculated the principal stresses, maximum shear stresses, and von Mises stresses, as shown in Fig. 13. Isolating the normal and shear stresses exclusively relating to holding pressure (P), it was observed that the maximum shear stress is greater than ones due to clamping force (C) or thermal loading (T). This data highlights the necessity to redefine the processing parameters in order to find lesser holding pressure values, which guarantee the quality of the moulded part. As it is a region responsible for the part conformation, mould design changes would also lead to a change in the part design. An important observation about this element is that it is undergoing positive normal stresses, even when only the stresses due to the holding pressure are considered. The animation of the C+P+T simulation was revised, confirming the behaviour of this critical element.

The stresses have exceeded the proportional limit in aluminium mould are unreal as the simulations do not consider elastic-plastic behaviour. The simulations only consider the linear mechanical properties of the materials, as shown in Tab. 1.

Description	Element ID = 1905864									
	Aluminium Mould Stresses (MPa)					Steel Mould Stresses (MPa)				
	C	P	T	C+P	C+P+T	C	P	T	C+P	C+P+T
Normal Stress XX (from solver)	-21.2	754.9	-76.5	733.6	657.2	-26.1	744.0	-154.1	717.9	563.8
Normal Stress YY (from solver)	-7.4	263.8	-29.2	256.5	227.2	-7.2	208.9	-48.2	201.8	153.6
Normal Stress ZZ (from solver)	-10.6	242.4	-34.1	231.8	197.7	-9.9	189.3	-50.7	179.4	128.7
Shear Stress XY (from solver)	8.7	144.7	-35.4	153.4	117.9	6.6	152.2	-49.1	158.8	109.7
Shear Stress YZ (from solver)	0.0	17.4	-5.5	17.4	11.9	-0.4	20.9	-7.5	20.5	13.0
Shear Stress ZX (from solver)	14.0	-151.7	9.6	-137.8	-128.2	16.8	-159.6	24.1	-142.7	-118.6
von Mises Stress (from solver)	31.1	-	-	607.0	538.4	36.0	-	-	645.4	507.9
Max Shear Stress (from solver)	17.5	-	-	327.0	288.6	19.9	-	-	347.5	272.3
Principal stress - 1 (σ1)	2.1	828.8	-7.0	807.0	715.8	1.9	819.8	-20.1	791.5	617.1
Principal stress - 2 (σ2)	-8.3	272.0	-36.8	261.9	227.7	-7.2	221.0	-56.4	211.1	156.4
Principal stress - 3 (σ3)	-32.9	160.3	-96.0	153.0	138.5	-37.8	101.4	-176.5	96.5	72.5
Max shear stress (τmax)	17.5	334.3	44.5	327.0	288.6	19.9	359.2	78.2	347.5	272.3
Von Mises stress (σv)	31.1	620.4	78.5	607.0	538.4	36.0	666.6	141.8	645.4	507.9

Fig. 13. Detailed stresses data of the critical mould element.

3.3 Parting Line Openings

Huang *et al.* [7] state that mould separation is often tolerated with an acceptable range (e.g., 0.075 mm) without causing flash. Therefore, as the cavity must not separate from the core more than 0.075 mm, it was observed that the aluminium mould tends to open more than the steel one, regardless the cooling condition. The maximum opening of the steel mould is 0.032 mm, while for the aluminium mould is 0.079 mm. Figure 14 shows the area where the maximum parting line openings were recorded, measured between the Cavity and the Core. Based on these results, solutions that aim to increase the aluminium moulds stiffness are necessary.

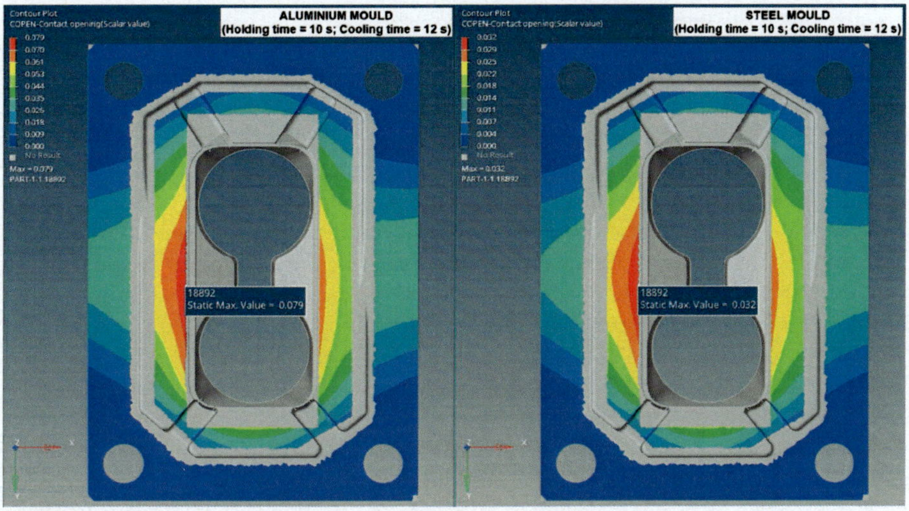

Fig. 14. Parting line opening during steady-state regime of a holding time of 10 s and cooling time of 12 s.

4 Conclusions

In this work, it was used a procedure based on finite elements to evaluate the temperature profile, the stress state and parting line openings in aluminium tooling, during the injection moulding process. The results were then compared with steel mould under the same operating conditions, which showed that the steady state in the steel mould occurs with a greater number of cycles, reaching higher temperature values than aluminium, but with lower thermal amplitude during the cycles. The maximum temperature is lower for longer injection cycles than for faster ones, with steel reaching higher temperatures than aluminium. The results show that the longer the injection cycle is, the closer the temperature profile of both moulds gets. It was noted that the average temperature of the moulded part when cooled in the aluminium mould for a period of 18 s is the same as the one cooled in a steel mould for 24 s, which proves the aluminium thermal efficiency compared to steel tooling. However, the time to reach the ejection, stablished by the temperature of the hot spot, is slightly different. To solve this issue, solutions involving cooling systems must be considered and it was not objective of this work.

Regarding the numerical stress results, it was collected for both moulds the maximum von Mises (failure criteria for steel mould) and shear (failure criteria for aluminium mould) stresses acting on the mould during the packing stage of the cyclic steady state regime, which is due to the temperature gradient, clamping force and holding pressure. The stress states observed are slightly greater for aluminium mould than for steel mould and becomes greater with the holding time increasing and lesser with the cooling time increasing. Additionally, to understand how the stress state behaves during the transient regime, maximum stresses were collected from all cycles of a specific case, and it was observed that stresses are greater for the steady state than for the first cycle.

The aluminium mould is susceptible to permanent deformations that steel is not, and changes in processing parameters or mould design are recommended so that aluminium benefits from its thermal advantages without harming the thermoplastic part manufacturing. The contribution of each loading was calculated, and the stresses caused by holding pressure were the greatest. The thermal loading relieved the stress state caused by clamping force and holding pressure. In addition, both the thermal loading and the clamping force decrease the stress state due to the holding pressure.

In relation to parting line openings, as the cavity must not separate from the core more than 0.075 mm to avoid flash, it was observed that the aluminium mould tends to open more than the steel one, regardless the cooling condition. The maximum opening for the steel mould is 0.032 mm, while for the aluminium mould is 0.079 mm. Based on these results, solutions that aim to increase the aluminium moulds stiffness are necessary.

Imperatively for this vehicular cupholder mould, the aluminium mould must be reinforced to reduce the parting line openings, and it is also important to use minimum holding pressure to minimize stresses. Therefore, the results showed that the difference between the mould's materials properties significantly changes the injection moulding conditions and the mould design, particularly the cooling system efficiency. These aspects are fundamental to make guidance to required modifications on the aluminium moulds, allowing the mould mechanical project and cooling system optimization to perform the best in terms of the part quality and process productivity. For future works, it is suggested to study different mould designs and different parameters combinations.

Acknowledgements. This work is a partial result from the Project DEMALAP - Demonstrador de Molde em Alumínio para Alta Produção (High production aluminium mold demonstrator) sponsored by Programa Rota 2030 from Brazilian Government described in Federal Law No. 13.755/2018 and it is coordinated by FUNDEP. The executors of this project are the universities SENAI CIMATEC, UFBA and UFSC. We thank the other partners of the project for their support: ESSS Engineering Simulation and Scientific Software Ltda.; Ford Motor Company Brasil Ltda., Moldit Brasil Ltda.; Open Mind Tecnologia Brasil Ltda.; Renault do Brasil S.A.; SMRC Fabricação e Comércio de Produtos Automotivos do Brasil Ltda. e Union Indústria de Moldes e Comércio Atacadista de Máquinas Eireli.

References

1. Farooque, R., Asjad, M., Rizvi, S.J.A.: A current state of art applied to injection moulding manufacturing process - A review. Mater. Today Proc. **43**, 441–446 (2020)
2. Arieta Filho, F.G. 2006. Ligas de alumínio de alta resistência para moldes de injeção de termoplásticos: ficção ou realidade? Revista Ferramental. Brazilian magazine of the tooling industry. https://issuu.com/revistaferramental8/docs/edi____o_8. Ed. 8. pp. 37-44. 19 Aug. 2022
3. Sardo, L., Daldoul, W., Vincent, M., Toulorge, T.: Simulations of heat transfer in thermoplastic injection molds manufactured by additive techniques. Int. Polym. Proc. **34**(1), 37–46 (2019)
4. Chen, S.-C., Liaw, W.-L., Su, P.-L., Chung, M.-H.: Investigation of mold plate separation in thin-wall injection molding. Adv. Polym. Technol. **22**(4), 306–319 (2003)
5. Zhou, T., Zhao, Y., Rao, Z.: Fundamental and estimation of thermal contact resistance between polymer matrix composites: a review. Int. J. Heat Mass Transf. **189**, 122701 (2022)

6. Somé, S.C., Delaunay, D., Faraj, J., Bailleul, J.L., Boyard, N., Quilliet, S.: Modeling of the thermal contact resistance time evolution at polymer-mold interface during injection molding: effect of polymers' solidification. Appl. Therm. Eng. **84**, 150–157 (2015)
7. Huang, M.-S., Nian, S.-C., Chen, J.-Y., Lin, C.-Y.: Influence of clamping force on tie-bar elongation, mold separation, and part dimensions in injection molding. Precis. Eng. **51**, 647–658 (2018)
8. Ahuett-Garza, H., Aguiar, F., Ponce, V.: Modeling and analysis of post ejection wall deflections in products made out of polypropylene. Polym. Eng. Sci. **50**(5), 1007–1018 (2010)

Check for updates

Numerical and Experimental Analysis of the Behavior of Tailor Welded Blanks to the Uniaxial Tensile Test

Guilherme Souza Assunção[1](✉), Etiene Andrade Pereira[1], Wellington Augusto dos Santos[1], and Alexandre Queiroz Bracarense[2]

[1] Federal University of Minas Gerais, Belo Horizonte, MG, Brazil
guilhermeassuncao@ufmg.br

[2] Pontifical Catholic University of Minas Gerais, Belo Horizonte, MG, Brazil

Abstract. In laser welding of Tailor Welded Blank (TWB), the melt zone (MZ) and heat-affected zone (HAZ) are very narrow, which makes it difficult to measure the mechanical properties of this region. This work sought to correlate the results of experimental uniaxial tensile tests with TWBs in which the weld line had different orientations, to numerical simulations. We worked with the Abaqus® software, considering the properties of the weld line, obtained with the Rule of Mixtures. TWB composed of interstitial-free steels of different mechanical properties was used, which led to the union of three different zones (two base materials and the welded region). The mechanical properties of the elastic and plastic regime of the tensile test were taken into account. It was found that weld makes a difference in the results, but for this to be really perceived, it is necessary to refine the mesh in the region, demanding more processing time. Whether or not the weld is considered, as long as it is qualified, rupture tends to occur in the material with the lowest mechanical resistance. The weld has greater mechanical strength and relative ductility, acting as a stress concentrator and changing the fracture direction of the TWB under tensile stress. The Rule of mixtures proved to be effective for representing the weld and the numerical model represented by the finite element method, with the imposed boundary conditions and simplification of the material as isotropic, proved to be valid with consistent results in relation to rupture load and elongation.

Keywords: Tailor Welded Blanks · Numerical simulation · Mechanical Characterization

1 Introduction

The manufacture of Tailor Welded Blanks (TWBs) involves joining materials with different thicknesses, properties and/or coatings, usually by laser welding (Marathe and Raval 2021; Andrade et al. 2019; Assunção et al. 2019). Despite the advantages of this technique, one of its biggest challenges is related to the fact that two different materials, with specific properties and thicknesses, will deform by conformation differently under the same loading conditions (Moayedi et al. 2020, Meng et al. 2020).

D. de Oliveira et al. (Eds.): COBEF 2023, LNME, pp. 70–82, 2024.
https://doi.org/10.1007/978-3-031-43555-3_7

One of the critical tests for characterizing the behavior under forming of TWBs are the tensile and limit forming curves (CLCs). Improving simulation can lead to increased failure prediction and greater assertiveness of tooling projects. In addition to reducing scrap rates, the correlation between experimental and simulated results increases from 80% to values greater than 90%, if there is a good selection of boundary conditions and the effects of the weld line are considered (Andrade et al. 2019, Moayedi et al. 2020, Meng et al. 2020 and Kardan et al. 2018).

However, according to Assunção et al. 2019, the difficulty of extracting the mechanical properties of the weld is highlighted, since the laser-welded region is very narrow, with this, conventional methods of analysis evaluate not only the weld, but also the materials base. Thus, processes such as the microhardness test and methods such as the Mixture Rule are used to minimize the influence of base materials on the mechanical characterization of the weld.

This work aims to study, analyze and develop solutions in computational parameters and boundary conditions for the characterization of the welded region of TWBs, from simulations by finite element method. The motivation is in the investigation of the critical points of design and simulation, as well as in the definition of the process parameters and in the selection of input and output data of the simulations, presenting a new detailing of the weld line in the numerical software to obtain a good correlation with the experimental results. (Aminzadeh et al. 2020 and Liavoli et al. 2020). The aim is to increase the correlation with mechanical tests, such as traction and CLC, of the plates made by TWB.

2 Methodology

In this section, the materials, equipment and procedures used to obtain the data and results of the work will be discussed. Figure 1 shows the flowchart of the work execution steps.

2.1 Materials

The plates used for analysis were provided, already laser welded, by FIAT Chrysler Automobiles (FCA) Brazil. The materials used in the manufacture of the plates are found respectively according to the codes CSN FEE210 and CSN FEP05, supplied by National Steel Company (CSN). Table 1 presents the mechanical properties, informed by CSN.

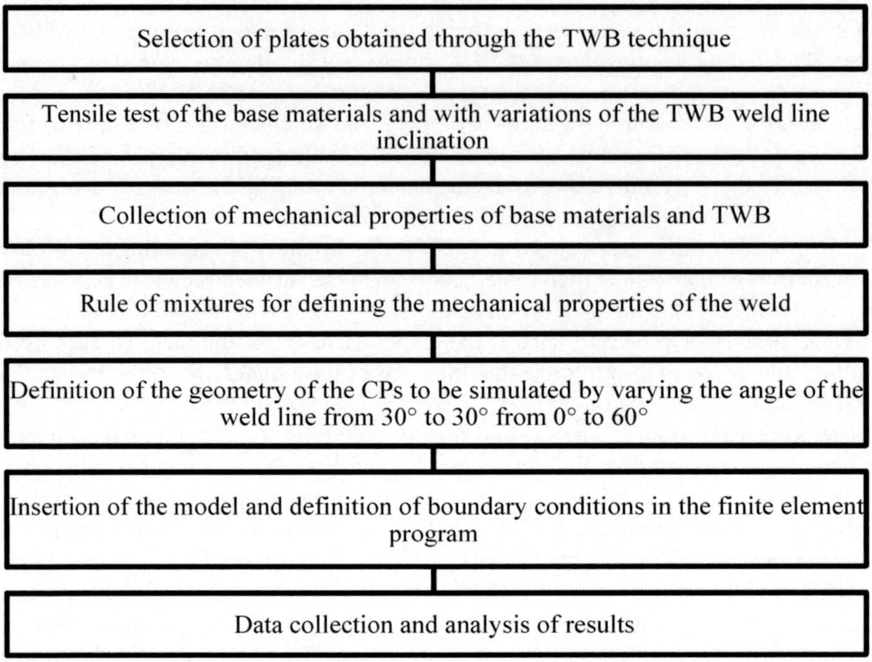

Fig. 1. Research execution steps.

Table 1. Chemical composition of IF steels used in TWB.

Element %	C	Mn	P	S	Si	N	Nb	Ti
IF FEP 05	0,0018	0,1350	0,0120	0,0100	0,0050	0,00420	0,0030	0,0570
IF FEE 210	0,0021	0,4320	0,0360	0,0120	0,0050	0,00360	0,0024	0,0260

Blanks are 0.65 mm and 1.1 mm thick, for materials FEP05 and FEE210, respectively, as shown in Fig. 2.

The welding process was by laser butt welding, aiming at the smallest possible extent for the Fused Zone (ZF) and Heat Affected Zone (HAZ). The CO2 SoundRonic® equipment was used with a nominal power of 8kW, with an average power of 65%, with pure helium as a protection gas with a flow rate of 10.5 L/min and a welding speed of 6.5 m/min.

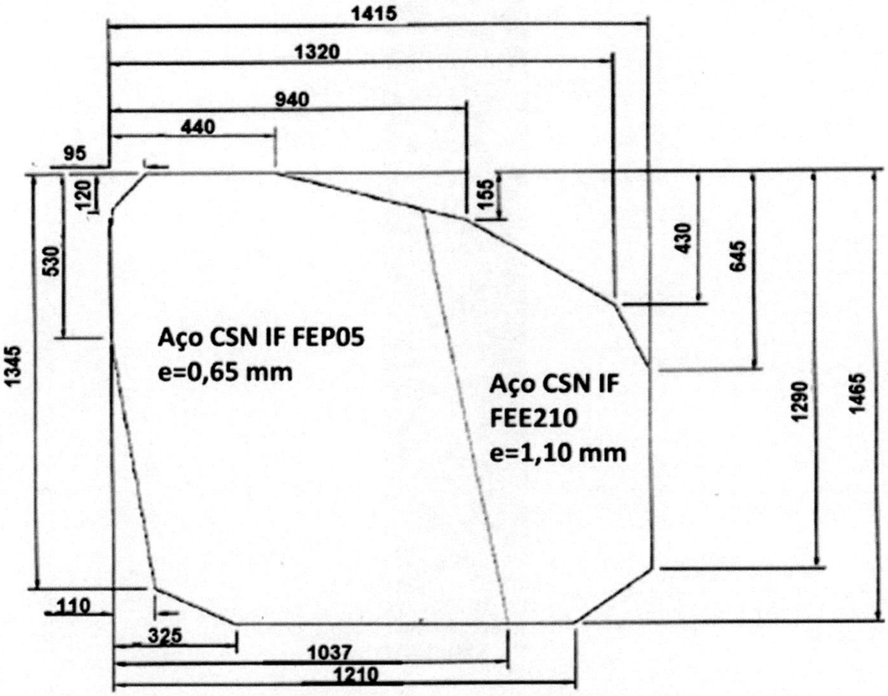

Fig. 2. Blank used to remove samples.

2.2 Tensile Test

In order to carry out the mechanical characterization of the base materials, and also of the weld line, tensile tests were carried out based on the ISO 6892–1 (2016) standard, working with triplicate for greater reliability and reduction of experimental errors. CPs were made both for the two base materials, FEP05 and FEE210, and for the TWB with the weld in the transverse of the entire specimen, as shown in Fig. 3.

An Instron® 3382 machine was used, equipped with a 2663–821 video extensometer and BlueHill®2 data acquisition software, with a head movement speed of 0.25 mm/s.

Three other models of specimens were tested to evaluate the slope of the weld line in the tensile test. 3 CPs of each type were evaluated, they are: 0°, 30° and 60° of inclination of the weld line.

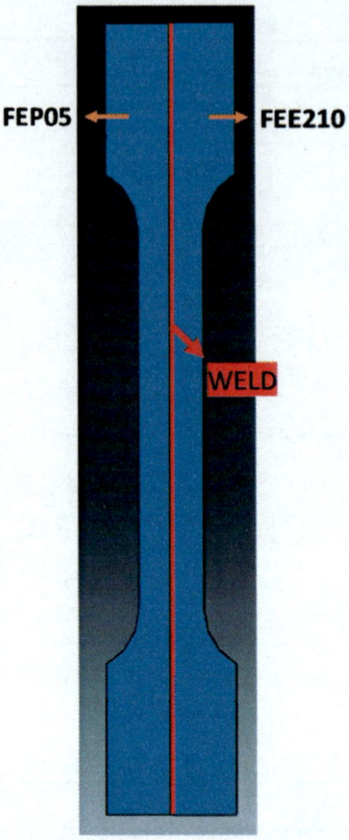

Fig. 3. Specimen for uniaxial tensile tests with the weld crossing the entire length.

2.3 Definitions and Boundary Conditions in the Finite Element Program

The definition of the model in the Abaqus® program starts with the creation of the three geometric parts that make up the tensile test of a TWB: thinner material (0.65 mm) with FEP05; thicker material (1.10 mm) with FEE210 and welded region that has variable geometry arising from the joining of the two metal sheets. The model was created in 3D space, with solid and deformable elements, created from extrusion, to represent the thickness of the plates.

The geometric pattern used for the tensile test was based on the ISO 6892–1 (2016) standard, as shown in Fig. 4.

The mesh determined for the CP is variable, in which the element size is 7 in the head of the specimen and 3.5 in the middle of the CP; as shown in Fig. 5. This variation was imposed to increase the refinement in the analysis region, reducing the number of CP nodes, increasing the reliability of the result and reducing the processing time. The element is of the hexahedral type, generated by scanning technique with medial axis algorithm.

As for the materials, the CP was divided into three distinct regions: FEP05, FEE210 and welded region (comprising ZTA and ZF together, for the purpose of simplifying the model). As common parameters, the steel density of 7.80 g/cm^3, isotropic materials, modulus of elasticity (E) of 210 GPa and Poisson coefficient (ν) of 0.30 were used, the latter two being useful to representation of the elastic phase of the material. For the

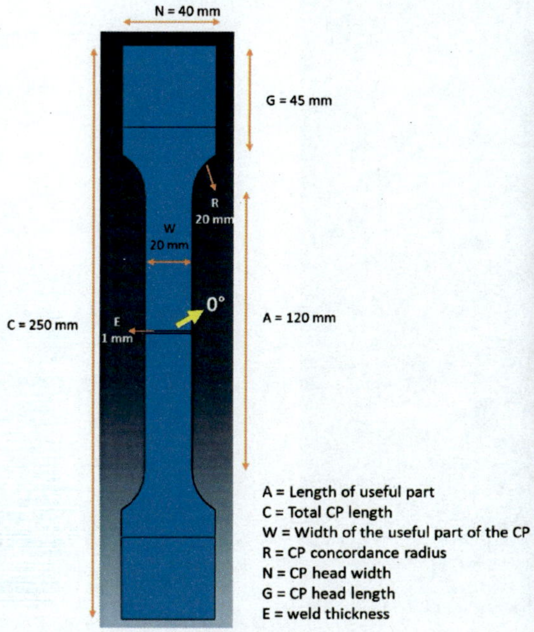

Fig. 4. Schematic representation of the dimensions of the specimens used in the uniaxial tensile tests with the weld inclined at 0°, 30° and 60°.

plastic regime, the Hollomon equation was used as presented in Eq. 1, in which the material strength coefficient (K) and hardening coefficient (n) data were based on the experimental tensile test.

$$\sigma = K.\varepsilon^{n} \tag{1}$$

In which:

- K = Resistance constant of the material.
- σ = Stress to which the material is subjected [MPa].
- Ɛ = Deformation to which the material is subjected.
- n = Work hardening exponent.

To represent the material in the welded region, the Rule of Mixtures, presented in Eq. 2, to define the tensile strength limit of the weld according to the weighting of the base materials and the welded region in a test specimen whose weld crosses the entire CP (Abdullah et al. 2001; Ciubotariu and Brabie 2011; Liavoli et al. 2020). The welded area was measured from images obtained by cutting the cross section and subsequent microscopic analysis and image processing in ImageJ® software.

$$\sigma_W = \frac{P - \left(K_1.\varepsilon_1^{n_1}\right).A_1 - \left(K_2.\varepsilon_2^{n_2}\right).A_2}{A_w} \tag{2}$$

In which:

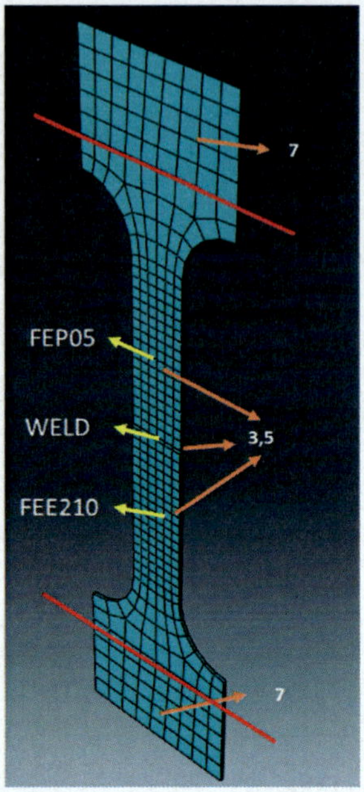

Fig. 5. Size of the mesh elements of the specimens used in the uniaxial tensile test

- σw = Resistance limit stress of the weld [MPa]
- P = Total load supported by the specimen [N]
- K1 = Coefficient of mechanical resistance of TWB material 1 [MPa]
- ε1 = True strain of material 1 of the TWB specimen
- A1 = Cross-sectional area of material 1 of the TWB specimen [mm^2]
- K2 = Coefficient of mechanical resistance of TWB material 2 [MPa]
- ε2 = True strain of TWB material 2
- A2 = Cross-sectional area of material 2 of the TWB specimen [MPa]
- Aw = Cross-sectional area of the weld of the TWB specimen [mm^2]

After creating the model and inserting the materials, the section of each part is determined, assigning the material in the solid category, of the homogeneous type, and the three bodies are joined (FEE05, FEE210 and weld).

When defining the incremental time (step), the type of procedure is used as explicit dynamic, with a period of 150 s with automatic increment. In mass dimensioning, it is defined as semiautomatic with a target time scale of 1 × 10–5, being applied right at the beginning of the test.

The output parameters of the simulation were stress, strain, displacement, and reaction forces. Among the stress, there is the main stress (S) and Von Mises (MISES), in the deformation there is the total deformation (E) and effective deformation (PEEQ), the displacements are calculated as a function of the displacement (U), velocity (V) and acceleration (A) and reaction forces (RF).

The contact properties between the bodies were defined as general contact between the parts (All with self), with tangential behavior without sliding (Rough).

Two restrictions were imposed to unite the three regions with the different materials, which are of the connection/node type, and which connect the material FEP05 to the weld and the FEE210 to the weld through the contact surfaces. And another constraint, of the kinematic coupling type, restricting all degrees of freedom, was placed on the head of the test specimen to carry out the simulation of the gripper in the tensile test.

Two boundary conditions were selected, the first being the displacement speed of the head of the traction test machine, applied to the reference point on the head of the specimen, positive direction of the Y direction (V2), with a speed of 0, 25 mm/s which represents the strain rate of 10–2 s-1. The second boundary condition is the locking of the CP base in the traction machine, therefore all translation and rotation movements in the lower part of the CP were restricted.

The analyzed results were registered as type of work with complete analysis and with double precision: analysis and data package.

3 Results

The first results obtained are from the tensile tests of the base materials and the specimen with the weld line crossing the CP. By manipulating the data and using Eq. (1) and Eq. (2), the mechanical properties of the two steels and the weld were obtained as shown in Table 2.

Table 2. Mechanical properties of FEE210, FEP05 and weld steels

	FEE210	FEP05	CP with weld 90°	Unit
Yield Stress (YS)	210,000	140,000	195,667	(MPa)
Tensile Strength Limit (LRT)	344,000	279,000	267,667	(MPa)
Engineering deformation (e)	0,215	0,240	0,200	-
True strain (Ɛ)	0,195	0,215	0,182	-
Strain-hardnening coefficient (n)	0,186	0,206	0,180	-
Coefficient of resistance (K)	466,357	382,891	393,955	(MPa)
Poisson coefficient (υ)	0,300	0,300	0,300	-
Modulus of elasticity (E)	2,10E+11	2,10E+11	2,10E+11	(Pa)
Weld area cp90	10,450	6,175	22,165	mm^2

With the definition of the mechanical properties of the base materials and the weld line, the numerical simulation of the tensile test was carried out in the finite element method program (Abaqus®), in which the slope of the weld line was varied at 0°, 30° and 60°. For comparison with the physical results, the breaking load and elongation data, presented in Fig. 6, were selected.

The breaking load showed similar behavior for both tests, in which the maximum load is approximately the same with the variation of the weld line inclination, as the CP tends to break in the material with less thickness and mechanical strength. These results were observed in other works such as those by Andrade et al. (2019) and Santos et al. (2021). This is valid as long as the weld does not have defects, as pointed out by Safdarian (2016), Fazli (2016) and Gautam et al. (2018). The difference between the

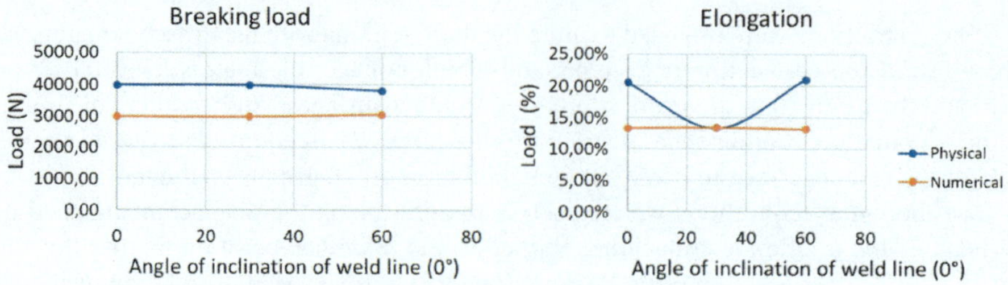

Fig. 6. Tensile load and elongation with varying weld line slope

physical and numerical test loads, 33.33%, may be related to the simplification of the model, which takes into account an isotropic material. However, it is known that IF steels have a high anisotropy coefficient (Hamidinejad et al. 2012).

To evaluate the total deformation as a function of the variation in the weld line inclination, the maximum elongation of the CPs was used. In the physical test, a similar behavior was obtained for angles of 0° and 60°, with elongations of 20.5% and 21% respectively. However, for 30°, the elongation of 13.33% indicates that at this angle of the weld line there was loss of elongation capacity, that is, less maximum deformation capacity, which indicates a lower conformation capacity in this configuration. This result is opposite to that presented by Andrade et al. (2019) and Santos et al. (2021), in which the 30° angle presents greater elongation and, consequently, would be a favorable angle for conformation. This phenomenon probably occurs because, although there is an attempt to find an optimal angle for the weld line orientation, there is no consensus on this issue, in addition to dealing with anisotropy, stress concentrators and other aspects of the base materials and the weld, mainly (Wang et al. 2021; Santos 2021).

In the numerical test, no relevant differences were observed for elongation as a function of the weld line inclination in the TWB, in which the maximum elongation was around 13% for all angles evaluated. This indicates that the elongation variation found in the physical test may be correlated with the anisotropy of the material and punctual points of weakness in the CPs, such as discontinuities in the weld line (lack of fusion, porosities, inclusions, filling failure or undercuts), region of removal of CPs or geometric irregularities of CPs (Ananthapadmanaban and Geethan 2020; Bagheri et al. 2020).

Regarding the fracture, Fig. 7 shows the three CPs ruptured in the physical and simulated tensile test, varying the inclination of the weld line.

It is observed that in the PC with a 0° inclination, in the physical sample there is a slight fracture inclination and in the numerical test the tendency is for the fracture to occur completely transversely to the PC. The transverse fracture is due to the homogeneous stress distribution on the weld line, as shown in Fig. 8 (A), where the weld is more stressed at both ends. Therefore, the trajectory followed in the physical test is related to the non-homogeneity of the material, causing the crack to propagate in a non-rectilinear direction (Arhumah, 2019; Sepasi and Ghadiri 2021; Wang et al. 2021).

For the 30° CP, equality is observed in relation to the fracture direction and the weld line inclination. In this case, the weld line begins to show greater stress in its upper portion, where there is a stress concentration seen in red in Fig. 8 (B), directly influencing the fracture angle, a phenomenon related to the shear planes of the material,

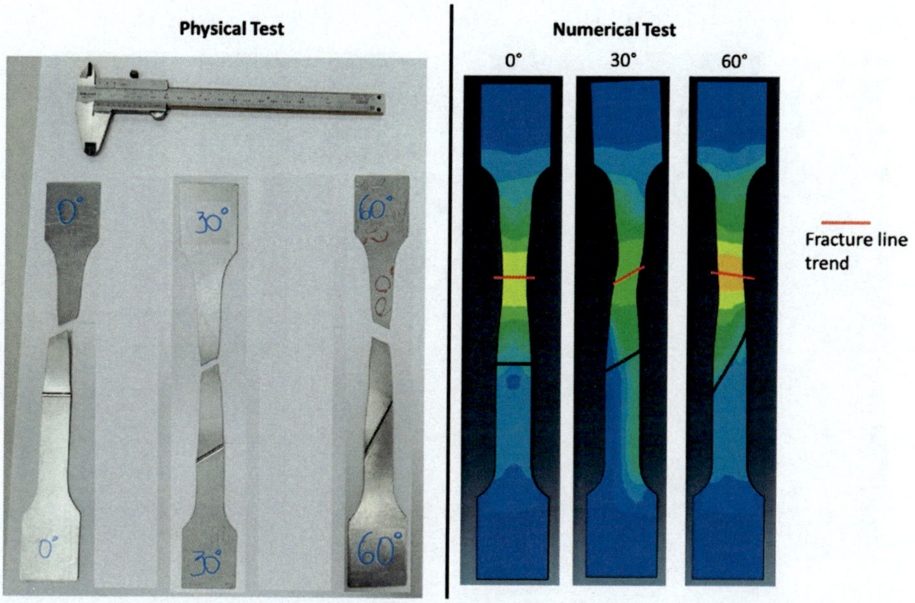

Fig. 7. Specimens (0°, 30° and 60°) broken in the uniaxial tensile test (A) Physical (B) Numerical

which determine the ductile fracture of IF steels, even if welded (Ananthapadmanaban and Geethan 2020).

With the weld inclination at 60°, the fractures for both tests occur in the same direction and angle, not following the direction of the weld line. The behavior of the weld fracture at 60° is justified by the greater restriction of movement of the weld line in its upper portion, generating a stress concentration in such a way that the beginning of the fracture occurs close to this region, highlighted in red in the Fig. 8 (C). This phenomenon, given the boundary conditions, simplifications, and constraints of the model, is explained by the fact that at the other end of the weld, in the lower region, there is even more material of lower mechanical resistance to deform and compensate for the tensile effort, changing the initial pattern. And fracture direction (Andrade et al. 2019).

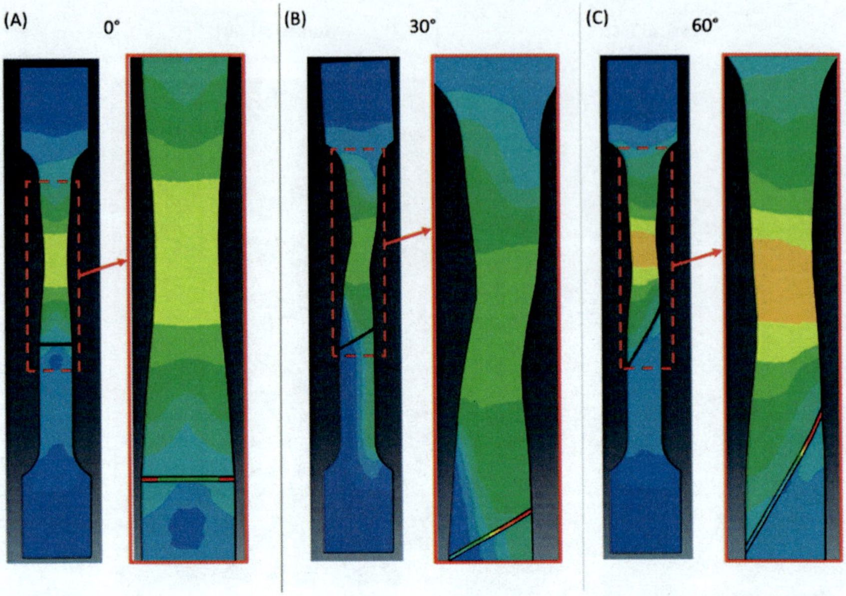

Fig. 8. Evaluation under tension of the weld line of the CPs (0°, 30° and 60°) in the numerical uniaxial tensile test

4 Conclusion

The present work analyzed a way to model and mechanically characterize the welded region of plates manufactured by Tailor Welded Blanks (TWB), by finite element method. By carrying out physical and numerical tensile tests on the base materials and TWB, it was possible to conclude that:

The use of the Rules of Mixtures proved to be effective to represent the mechanical properties of the weld. To make the representation more faithful to reality, the use of sub-size specimens is indicated, bringing the properties of the weld line even closer.
The numerical model represented by the finite element method proved to be valid with results consistent with the imposed boundary conditions. The differences observed in the comparison between the physical and numerical tests of the breaking load and elongation may be linked to the anisotropy of the steels free of interstitials.
Regarding PCs fracture, it was observed that for 0° the fracture in the physical test should occur transversely if the material was homogeneous and without defects. For 30° and 60°, similar results were obtained in the comparison between the physical and numerical, in which the stress concentration generated in the weld line changes according to the variation of the inclination angle, causing it to break with parallel angulation the direction from 30° and opposite direction to 60°.

References

Abdullah, K., Wild, P.M., Jeswiet, J.J., Ghasempoor, A.: Tensile testing for weld deformation properties in similar gage tailor welded blanks using the rule of mixtures. J. Mater. Process. Technol. **112**(1), 91–97 (2001). https://doi.org/10.1016/S0924-0136(01)00555-6

Aminzadeh, A., Parvizi, A., Moradi, M.: Multi-objective topology optimization of deep drawing dissimilar tailor laser welded blanks, experimental and finite element investigation. Optics Laser Technol. **125**, 106029 (2020)

Ananthapadmanaban, D., Geethan, K., Arun, V.: Fracture behavior of solid-state welded joints. In: Thanapalan, K. (ed.). Engineering Failure Analysis. Cap. 5, p. 17. Intechopen (2020)

Andrade, E.P.: Mechanical characterization and microstructural analysis of plates obtained by the Tailor Welded Blank (TWB) process. 2019. Dissertation (master's in mechanical engineering) - Federal University of Minas Gerais, Belo Horizonte (2019)

Andrade, E.P., Assunção, G.S., Santos, W.A., Bracarense, A.Q., Felizardo, I.: Mechanical characterization and microstructural analysis of plates obtained by the tailor welded blank (TWB) process. Weld. Inspect. **24**, e2425 (2019)

Arhumah, Z.: Characterization of microstructure and mechanical behaviour of heat affected Zones in robotic arc welding of AA6061-t6. 2019. 133 f. Dissertation (Master's in mechanical engineering) École de Technologie Supérieure Université Du Québec, Montreal (2019)

Assunção, G.S., Andrade, E.P., Santos, W.A., Felizardo, I., Bracarense, A.Q.: Mechanical characterization of the welded region of tailor welded blanks (TWB) based on the microhardness profile. Weld. Inspect. **24**, 1–10 (2019)

Brazilian Association of Technical Standards (ABNT) – NBR-ISO – 6892–1. Metallic Materials - Tensile Test at Ambient Temperature

Bagheri, B., Abbasi, M., Hamzeloo, R.: Comparison of different welding methods on mechanical properties and formability behaviors of tailor welded blanks (TWB) made from AA6061 alloys. Proc. Inst. Mech. Eng. Part C. J. Mech. Eng. Sci. **235**(12), 2225–2237 (2020)

Ciubotariu, V., Brabie, G.: Weld line behaviour during uniaxial tensile testing of tailor welded blanks. Arch. Civil Mech. Eng. **11**(4), 811–824 (2011). https://doi.org/10.1016/S1644-9665(12)60079-X

Fazli, A.: Investigation of the effects of process parameters on the welding line movement in deep drawing of tailor welded blanks. Int. J. Adv. Des. Manuf. Technol. **9**(2), 45–52 (2016)

Gautam, V., Raut, V.M., Kumar, D.R.: Analytical prediction of spring back in bending of tailor-welded blanks incorporating effect of anisotropy and weld zone properties. Proc. Inst. Mech. Eng. Part L. J. Mater. Des. Appl. **232**(4), 294–306 (2018)

Hamidinejad, S.M., Hasanniya, M.H., NooshinSalari, E., Valizadeh, E.: CO2 laser welding of interstitial free galvanized steel sheets used in tailor welded blanks. Int. J. Adv. Manuf. Technol. **64**(1–4), 195–206 (2012). https://doi.org/10.1007/s00170-012-4005-7

Kardan, M., Parvizi, A., Askari, A.: Influence of process parameters on residual stresses in deep-drawing process with FEM and experimental evaluations. J. Braz. Soc. Mech. Sci. Eng. **40**(3), 1–12 (2018). https://doi.org/10.1007/s40430-018-1085-9

Liavoli, R.P., Gorji, H., Bakhshi-Jooybari, M., Mirnia, M.J.: Investigation on formability of tailor-welded blanks in incremental forming. Int. J. Eng. **33**(5), 906–9115 (2020)

Marathe, S., Raval, H.: Investigation on effect of weld yield strength during single point incremental forming of tailor welded blanks. Int. J. Mod. Manuf. Technol. **13**(1), 123–131 (2021)

Meng, D., Zhang, L., Li, Y., Zhang, G.: Welding deformation of unequal thickness tailor welded blanks under multiple factors. IoP Conf. Ser. Earth Environ. Sci. **508**(1), 012223 (2020)

Moayedi, H., Darabi, R., Ghabussi, A., Habibi, M., Foong, L.K.: Weld orientation effects on the formability of tailor welded thin steel sheets. Thin-Walled Struct. **149**, 106669 (2020)

Safdarian, R.: The effects of strength ratio on the forming limit diagram of tailor-welded blanks. Ironmaking Steelmaking. **45**(1), 17–24 (2016)

Sepasi, M., Ghadiri, F.: Application of experimental investigation of forming limit diagrams of tailor welded blanks in ship hull design. Iranian PMO - Ports and Maritime Organization (2021)

Wang, H., Liu, L., Wang, H., Zhou, J.: Control of defects in deep drawing of tailor-welded blanks for complex shape automotive panel. J. Adv. Manuf. Technol. 1–20, Research Square (2021)

Microstructure and Hardness of S41003 Steel When Submitted to Different Heat Treatments

Renata Rodrigues Alves[1(✉)], José Lucio Gonçalves Junior[1], Daniel Fernandes Cunha[1], Kléber Mendes de Figueiredo[1], Geraldo Lucio De Faria[2], and Wilian da Silva Labiapari[3]

[1] Federal University of Goiás, Goiânia, GO, Brazil
renataufg1@gmail.com
[2] Federal University of Ouro Preto, Ouro Preto, MG, Brazil
[3] Aperam South America Research Center, Timóteo, MG, Brasil

Abstract. Although the ferritic stainless steel 410D (S41003) has a relatively simple chemical composition, with only 0.015% carbon by weight, previous studies indicate that this alloy can be fully or partially hardened by heat treatment. Therefore, the present work aims to analyze the effect of different heat treatment conditions on the microstructure and hardening of the material. As a methodology, two heat treatment conditions were performed. In the first condition, the samples were heated to 1000 °C, varying the austenitizing time at 5 min, 30 min, 60 min, 120 min, and 240 min and then quenched in water. For the second condition, the samples underwent an intercritical annealing at 800 °C for 1 h and were cooled in the furnace. Afterwards, the samples were heated to 1000 °C, varying the austenitizing time at 5 min, 30 min, 60 min, 120 min, and 240 min and quenched in water. The microstructural analysis indicated that the as-received material has a polygonal ferritic structure with $Cr_{23}C_6$ carbide precipitation, and the annealing treatment generated the precipitation of more chromium carbides. After the quenching treatment, the qualitative analysis corroborated previous studies indicating an increase in lath martensite due to the increase in grain size. The hardening results indicate that the alloy can be hardened by quenching, but when the material undergoes a prior intercritical annealing process, the hardness increases. Moreover, the longer the samples are held at the austenitizing temperature, the lower the post-quenching hardness, a phenomenon associated with the increase in austenitic grain size.

Keyword: Stainless steel · Microstructural characterization · Hardness · Grain size · Chromium carbide

1 Introduction

Ferritic stainless steels are iron-chromium-carbon alloys, with Cr in the range of 12 to 30 wt%, and a C content lower than 0.1 wt%. They have an α ferrite phase and a body-centered cubic (BCC) crystal structure. They offer good corrosion resistance, are magnetic, and their mechanical behavior is similar to carbon steel [1, 2].

In addition to the basic Fe-Cr composition, other elements are added to stainless steels to interfere with the mechanical properties and microstructure of the material. Regarding microstructure, the added alloying elements influence not only changes in equilibrium

D. de Oliveira et al. (Eds.): COBEF 2023, LNME, pp. 83–94, 2024.
https://doi.org/10.1007/978-3-031-43555-3_8

phases or constituents but also in the way and speed at which the phases form. Austenite-stabilizing elements, for example, tend to decrease the temperature of transformation from the γ austenite phase to the α ferrite phase and increase the temperature from the γ austenite to the σ ferrite phase. Nickel and manganese are elements that, when added in high concentration, act as austenite stabilizers [3]. On the other hand, chromium, molybdenum, and niobium stabilize the ferrite phase, decreasing the range where the austenite phase exists [4].

In theory, ferritic stainless steels cannot be fully austenitized and, as a result, do not exhibit significant hardening through heat treatment. However, there are exceptions, such as the ferritic stainless steel 410D, which, due to its low chromium content and other ferrite-stabilizing elements can be fully or partially hardened by an appropriate heat treatment [5–7]. When partially hardened, at the intercritical temperature, the material forms a two-phase structure composed of ferrite and martensite. Martensite is a harder and more resistant phase, while ferrite is soft, generating a combination of properties that may be favorable depending on the situation [8].

Ferritic stainless steel 410D, also known as UNS S41003 or 3Cr12, has structural and wear applications. The alloy contains low carbon content with additions of chromium and nickel, being used in environments that require resistance to corrosion and abrasion. Due to the relatively simple chemical composition, this stainless steel is an option for replacing materials, such as carbon steel [9].

The 410D alloy is susceptible to the precipitation of chromium carbide $(Cr, Fe)_{23}C_6$, after its manufacturing process, with an average size of 500 nm and generally grouped in bands along the rolling direction [10]. For the alloy with 0.014 wt% C and 11.06 wt% Cr, studied by researchers Moura et al. [11], after hot rolling, the material presented a ferritic grain size of (22.1 ± 2.6) μm with (0.005 ± 0.001) % area of $(Cr, Fe)_{23}C_6$. The same material, after annealing at 795 °C for 5 h, presented a ferritic grain size of (23 ± 3) μm with (0.05 ± 0.001) % area of $(Cr, Fe)_{23}C_6$, therefore, although the grain size did not vary significantly with annealing, the carbides fraction increased.

Faria et al. [12], in their work with 410D alloy, containing 0.013% and 11.097% weight of carbon and chromium, respectively, investigated the development of microstructure and mechanical properties when subjected to intercritical temperatures of 825 °C and 850 °C for different times and water quenched. The results demonstrated a relationship between time, temperature, and the volumetric fraction of martensite, showing an increase in martensite formation accompanied by a decrease in chromium carbide precipitation. Experimental observations suggest that the dissolution of $Cr_{23}C_6$, chromium homogenization, and the presence of carbon in solid solution significantly influence the kinetics of austenitization and subsequent transformation into martensite during the quenching process [10, 13].

Another known factor that influences martensite transformation is the austenite grain size. The grain boundary provides a favorable environment for martensite nucleation, but the grain size has a stabilizing effect, limiting the volume of martensite [14].

Previous studies have discussed the influence of temperature and time on the microstructure and strength of 410D steel after the quenching process. However, as far as the authors knowledge, no detailed investigation has been conducted regarding the comparison between quenching process with previous intercritical annealing. The results obtained in this study provide valuable insights for enhancing the control of material properties.

2 Materials and Methods

2.1 Materials

The material studied is the ferritic stainless steel 410D, hot rolled and annealed, produced by the company Aperam South America. The chemical composition of the steel is presented in Table 1. The Carbon and Silicon were analyzed by the infrared absorption technique on the manufacturer's equipment Leco, model CS444Ò, while the Nitrogen content was evaluated by the thermocouple technique on the manufacturer's equipment Leco, model TC436Ò. The remaining chemical elements were analyzed by the x-ray fluorescence spectrometry technique on the manufacturer's equipment ThermoARL, model 9900.

Table 1. Chemical composition of 410D (wt. %).

Element	C	Mn	Si	P	S	Cr	Ni
Obtained result	0,0150	0,58	0,50	0,028	0,0008	11,16	0,34

According to ASTM A240/A240M [15] standard, steel 410D is classified as ASTM S41003. To identify the phase transformation temperatures and precipitates present in the material, a computational simulation was performed using the Matcalc software, database mc-fe-v2.060, based on the steel's chemical composition.

The equilibrium phase diagram, Fig. 1, indicates that the initial austenitization temperature is A_c1 = 778 °C and the final austenitization temperature is A_c3 = 843 °C.

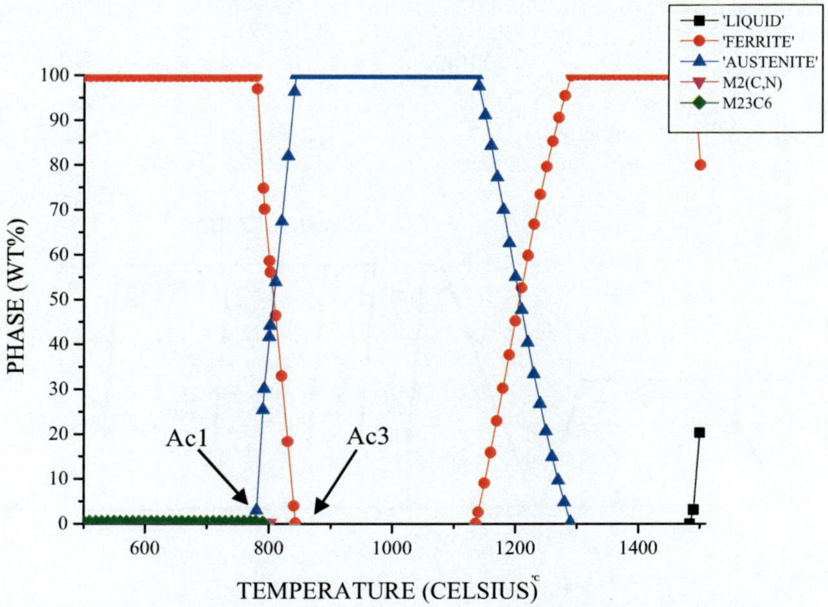

Fig. 1. The computational simulation yielded the phase diagram for the 410D steel.

2.2 Experimental Procedures

The 410D stainless steel samples have a thickness of 5.4 mm and dimensions of 25.3 × 20.8 mm. Two heat treatment conditions were performed: in the first condition, the material was quenched by continuously heating it up to 1000 °C, where it was completely austenitized, and then cooled in water. The austenitization time varied as shown in Fig. 2. In the second condition, a prior intercritical annealing was performed at 800 °C for 1 h followed by slow cooling inside the furnace, before quenching.

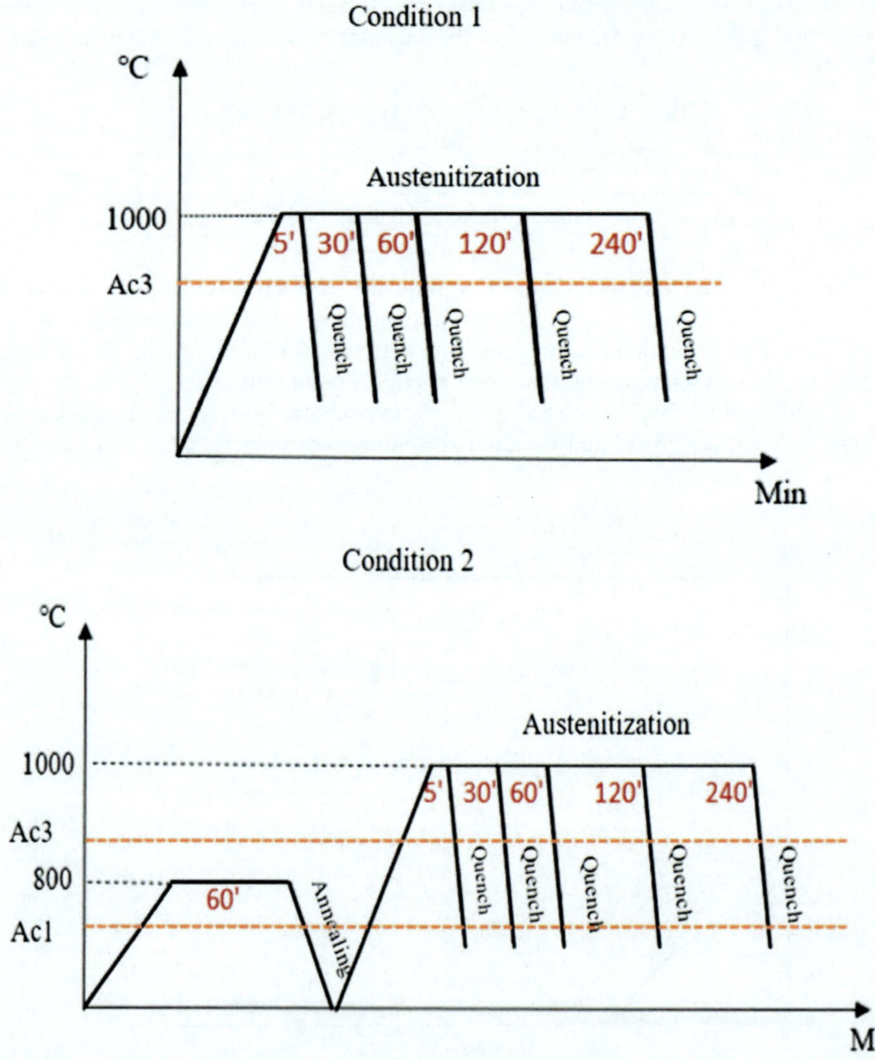

Fig. 2. Experimental procedure diagram.

2.3 Hardness Test

The Vickers hardness test was performed according to the ASTM E 92-82 [16] standard, using a Mitutoyo HM-100 hardness tester. Ten indentations were performed using a 10 kgf load, with each indentation lasting for over 10 s.

2.4 Microstructural Characterization

The samples, as-received and after heat treatments, were prepared metallographically with water sandpaper of 220, 320, 400, 500, 600, and 2000 grit; then they were polished with alumina suspension of 1 μm and 0.5 μm grit size. For chemical etching, the reagent was selected according to ASTM E407-07 [17] standard, using solution 81 (Vilella) with the chemical composition presented in Table 2. For microscopic analysis, the ZEISS optical microscope, model AXIO Imager M2m, and the JEOL scanning electron microscope, model JSM IT300LV, using a tungsten filament electron beam were used.

Table 2. Chemical etching performed according to ASTM E407-07 [17] standards.

Villela Composition	Application Method	Etching Time
1 g picric acid 100 mL ethanol (95%) 5 mL HCl	Immersion	40 s

To assess the chromium concentration in the samples, as received and after heat treatment, composition maps were obtained using Electron Probe Microanalyzers (EPMA). The samples were analyzed in a JEOL JXA-8230 electron microprobe, with TAP, PET/L-H, and LIF-L/H crystals. The analyses were performed with an acceleration voltage of 15 kV and a current of 500 nA.

2.5 Grain Size

To reveal the grain boundaries, several attempts with different etchants were performed, and the composition that showed the best result is presented in Table 3.

Table 3. The chemical etchant used to reveal the grain boundaries.

Composition	Application Method	Etching Time
• 15 ml dishwashing detergent • 20 ml Nital 5% (5 ml HNO3 + 95 Ethyl Alcohol) • 60 ml Villela	Immersion	45s

The grain size was determined using the ImageJ computer program, where a random selection of 40 grains was made in each sample for calculating the arithmetic mean and standard deviation.

3 Results and Discussion

3.1 Effect of Annealing on Microstructure

Figure 3 illustrates the microstructure of the 410D steel in both its as-received and annealed states. The presence of polygonal ferrite grains can be observed, with a grain size of (26 ± 6) μm in the as-received state and (23 ± 6) μm after annealing. As demonstrated by Bettanini et al. [10], carbides appear in bands along the rolling direction within the ferrite matrix.

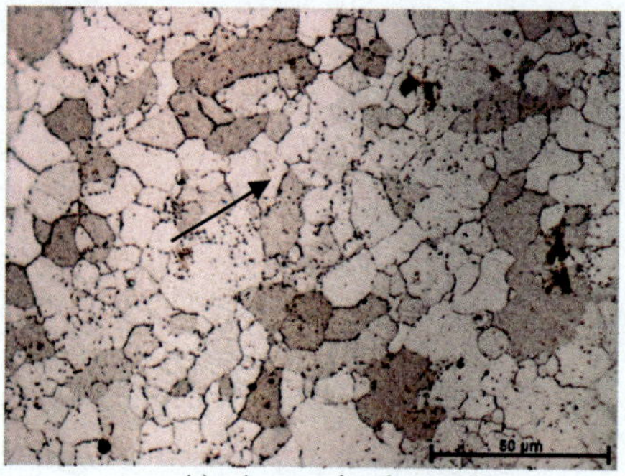

A) As-received sample.

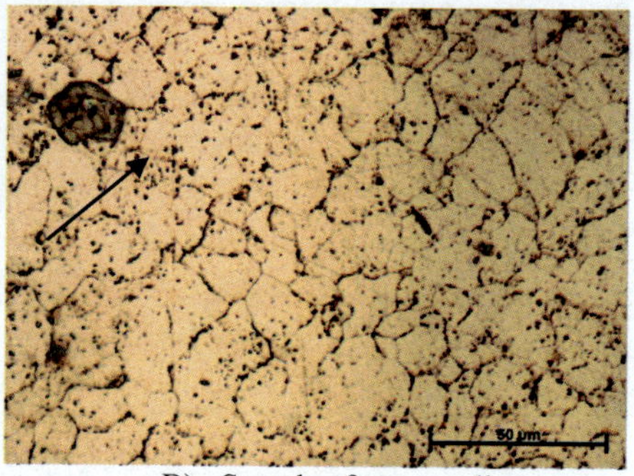

B) Sample after annealing.

Fig. 3. Micrographs exhibit the distribution of $M_{23}C_6$ carbides within the ferrite matrix, forming bands along the rolling direction, indicated by the arrow.

As shown in the equilibrium diagram, Fig. 1, the 410D steel presents $M_{23}C_6$ type carbides up to a temperature of around 800 °C. Figure 4 displays the micrograph of the as-received sample, revealing a spheroidized structure of the precipitates primarily located at the grain boundary, which is a region prone to higher energy accumulation.

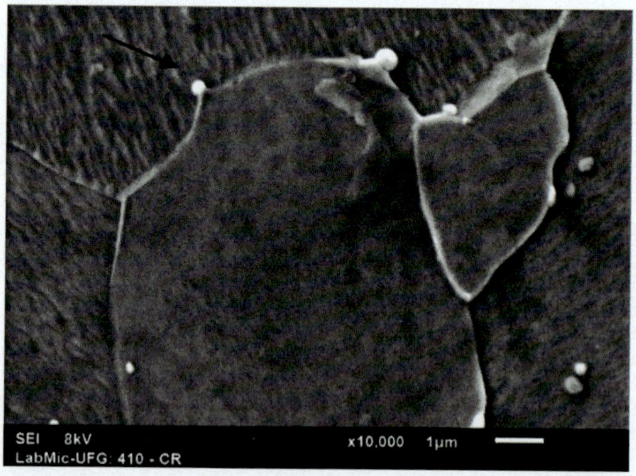

Fig. 4. Microstructure of as-received 410D steel, with carbide indicated by arrow.

Carbon is an interstitial element with low solubility in Fe α and when combined with chromium, forms the precipitate $Cr_{23}C_6$, generating regions poor in chromium around it (Serna-Giraldo, 2007). Therefore, in order to investigate the distribution of $Cr_{23}C_6$ carbides, chromium concentration maps were obtained, as shown in Fig. 5.

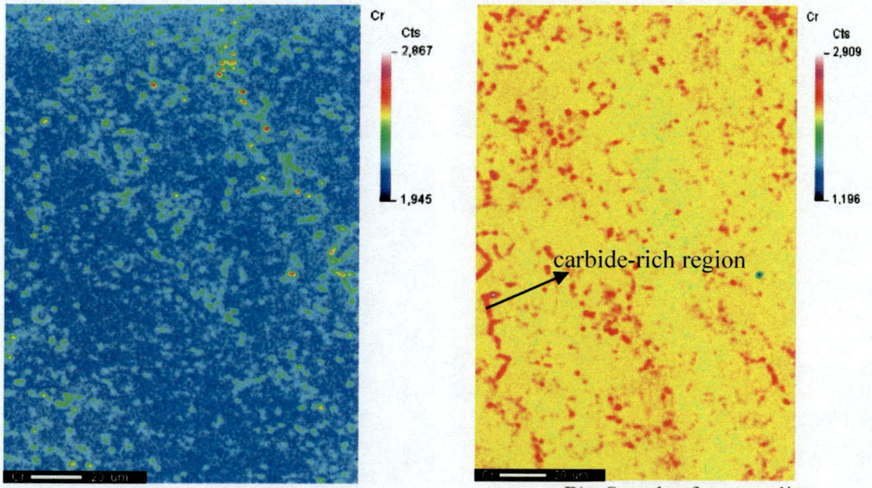

A) As-received sample. B) Sample after annealing.

Fig. 5. Chromium concentration maps in the 410D steel sample before and after annealing at 800 °C.

The color scale of the map indicates that the sample annealed at 800 °C, Fig. 5(B), has more chromium-enriched regions due to the precipitation of $Cr_{23}C_6$, compared to the as-received sample, Fig. 5(A). This result is consistent with the findings of Moura et al. [11], suggesting a higher fraction of chromium carbide in the annealed sample.

3.2 Effect of Austenitization Time in Microstructure

The micrograph of the samples in condition 1 and of the samples in condition 2 show similar results and discussions. Therefore, in this section, images of condition 1 will be presented, and the discussion regarding the evolution of the microstructure can be extended to condition 2.

Observing the equilibrium diagram, it can be seen that at 1000 °C, the 410D steel is completely austenitized, therefore, martensite formation occurs after quenching. With the increase in austenitization time, the martensite lath becomes more perceptible, as can be observed in Fig. 6(A), (B), (C), (D), and (E).

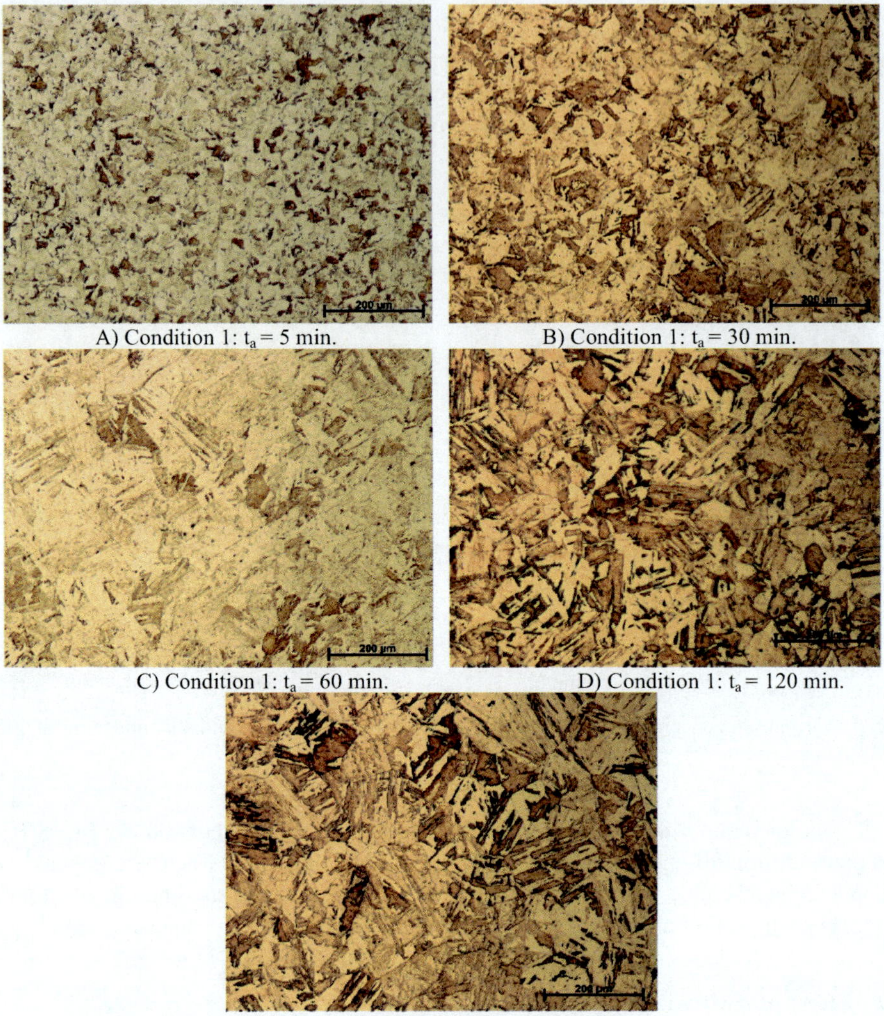

A) Condition 1: $t_a = 5$ min.

B) Condition 1: $t_a = 30$ min.

C) Condition 1: $t_a = 60$ min.

D) Condition 1: $t_a = 120$ min.

E) Condition 1: $t_a = 240$ min.

Fig. 6. Microstructure after condition 1 heat treatment, for each austenitization time (t_a).

The analysis of Fig. 6 points to an increase in the thickness of the martensite with an increase in the austenitizing time, suggesting that an increase in the prior austenitic grain size is the underlying cause of this phenomenon. The size of the prior austenitic grain limits the size of the martensite, for small grains the laths collide and stop growing earlier and therefore become thinner and shorter [18]. Verdiere et al. [19] also concluded that an increase in the austenitizing time results in a significant enlargement of the grain size, leading to larger and wider martensite laths due to the increased block size.

In the quenching process, with an austenitizing time of 5 min, it was not possible to observe the microstructure by optical microscopy, Fig. 6(A), making it necessary to use scanning electron microscopy to identify the martensitic phase, Fig. 7.

Fig. 7. The microstructure following the heat treatment in condition 1, heated to 1000 °C for 5 min (A), followed by water quenching.

In addition, although the equilibrium diagram indicates that the carbides would be completely austenitized at 1000 °C, this process is relatively slow and a gradual dissolution of the carbides was observed with increasing austenitization time, as identified in Fig. 8.

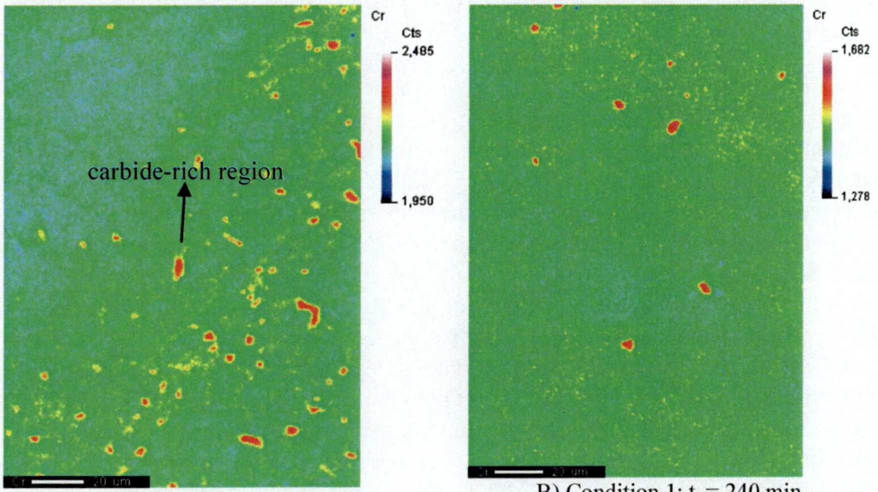

A) Condition 1: ta = 5 min.

B) Condition 1: t_a = 240 min.

Fig. 8. Chromium concentration maps, obtained by electron probe microanalysis, in samples of 410D steel heated at 1000 °C for 5 min (A) and 240 min (B), after quenched in water.

3.3 Effect of Austenitization Time on Hardness and Grain Size

Both samples from Condition 1 and Condition 2 were hardened by the heat treatment of quenching, as shown in Fig. 9. However, the samples that underwent intercritical annealing prior to quenching, Condition 2, exhibited higher hardness than the samples only quenched, Condition 1.

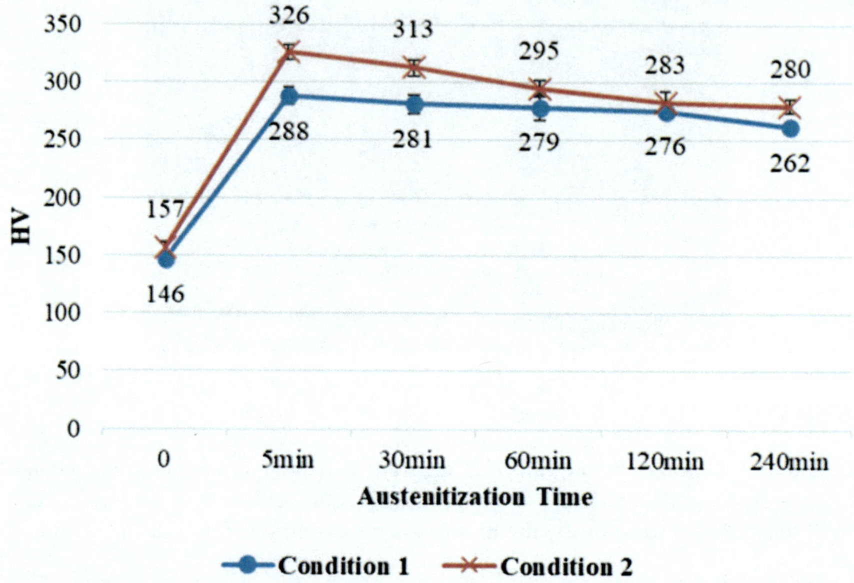

Fig. 9. Result of Vickers hardness test on 410D steel, tested with a load of 10 Kgf for 10 s.

The increase in hardness due to intercritical annealing can be justified by the increase in the fraction of $Cr_{23}C_6$, as indicated in the map of Fig. 5. Chromium carbide is a nanoparticle that can serve as an obstacle to dislocation movement and prevent grain boundary sliding, thus hardening the material [20, 21]. Another intriguing observation is that as the austenitization time increases, the disparity in hardness between Condition 1 and Condition 2 diminishes, providing support for the hypothesis regarding the influence of carbides, since it dissolves with increasing austenitization time.

The hardness test conducted on both conditions also reveals a reduction in hardness as the austenitization time increases. The likely explanation is based on the increase in prior austenitic grain size, as shown in Fig. 10. The increase in grain size is correlated with a reduction in grain boundary density. Grain boundaries serve as obstacles to dislocation movement due to the mismatch in crystallographic orientations between adjacent grains, impeding their motion. Moreover, the presence of atomic disorder in the grain boundary region leads to a disruption of the slip planes. Consequently, a material with a finer grain size, characterized by smaller grains, exhibits a larger collective grain boundary area. This increased boundary area hinders dislocation movement, resulting in enhanced material hardness [22].

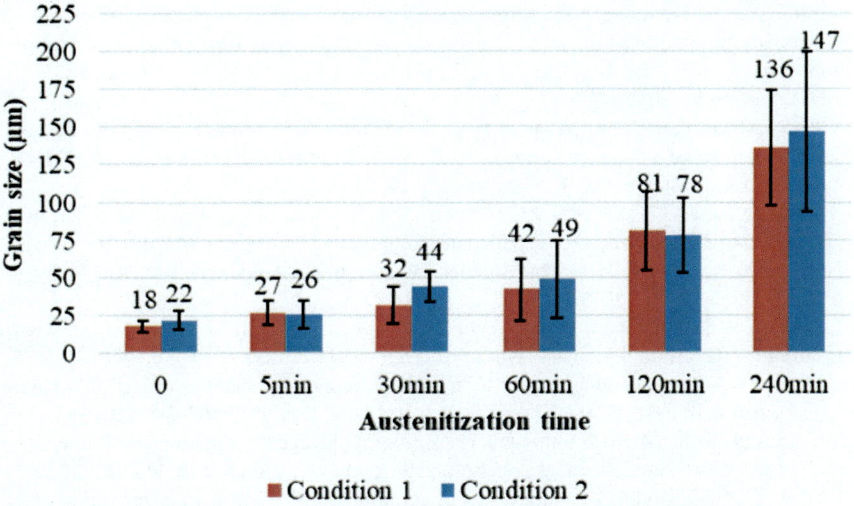

Fig. 10. Pre-existing austenitic grain size, in micrometers, of 410D steel samples.

4 Conclusion

Based on the evidence produced in this research, according to the analysis and discussion of the results presented here, the following conclusions are presented:

In terms of microstructure, it was observed that the material subjected to intercritical annealing showed minimal variation in ferritic grain size, while experiencing an increase in chromium carbide precipitation. The micrographs following the quenching process, in both conditions, reveal an enlargement in the width and size of martensite with an extended austenitization time.

This study also observed that the 410D steel can be hardened when austenitized and water quenched. Condition 1 had a maximum hardness increase of 97%, while Condition 2 had 108%. The higher hardness observed in Condition 2 can be attributed to the greater presence of chromium carbide in the samples annealed at an intercritical temperature. However, as the austenitization time increased, the samples exhibited softening, and the reduction in hardness is linked to a decrease in the overall grain boundary area. This decrease facilitates dislocation movement of dislocations.

References

1. Chiaverini, V.: Aços e ferros fundidos, 7th edn. ABM, São Paulo (1999)
2. Modenesi, P.J.: Soldabilidade dos Aços Inoxidáveis. SENAI-SP, São Paulo (2001)
3. Silva, A.L.V.C., Mei, P.R.: Aços e ligas especiais, 3rd edn. Edgard Blucher, São Paulo (2010)
4. Boniardi, M., Casaroli, A.: Stainless Steels, 1st edn. Gruppo Lucefin, Brescia (2014)
5. Sundqvist, J., Mannien, T., Heikkinen, H.P., Anttila, S., Kaplan, A.F.H.: Laser surface hardening of 11 % Cr ferritic stainless steel and its sensitisation behaviour. Surf. Coatings Technol. **344**, 673–679 (2018)

6. Arruda, A.A.F., Carvalho, A.C., Moreira, P.S., Queiroz, R.R.U., Faria, G.L.: Características da camada de óxidos formada no aço inoxidável 410D após tratamentos isotérmicos em atmosfera oxidante. In: Blucher, 73º Congresso Anual da ABM, vol. 73, pp. 1470–1481. ABM, São Paulo (2018)
7. Vilela, L.B.M., Faria, G.L., Alcantara, C.M., Oliveira, T.S., Cota, A.B.: Efeito da taxa de resfriamento sobre a formação de martensita em um aço inoxidável ferrítico com 11 %Cr e baixos teores de intersticiais. Revista Materia **24**, 1–11 (2019)
8. Nunes, I.P., Faria, G.L., Lacerda, J.C., Godefroid, L.B.: Influência de Tratamentos Térmicos de Têmpera com Austenitizaçao Intercrítica no Comportamento Mecânico de um Aço Inoxidável Ferrítico do Tipo UNS S41003. In: Blucher, 74◦ Congresso Anual da ABM, vol. 74, pp. 440–451. ABM, São Paulo (2019)
9. Labiapari, W.S., Alcântara, C.M., Costa, H., Mello, J.D.B.: Stainless steel as an antiwear material for the bio-fuel industry. Wear **302**, 1536–1545 (2013)
10. Bettanini, A.M., et al.: Influence of M23C6 dissolution on the kinetics of ferrite to austenite transformation in Fe-11Cr-0.06C stainless steel. Mater. Des. **162**, 362–347 (2019)
11. Moura, A.N., et al.: Microstructure and crystallographic texture of martensitic stainless steels after cold rolling and subsequent recrystallization. Mater. Charact. **192**, 112190 (2022)
12. Faria, G.L., Godefroid, L.B., Nunes, I.P., Lacerda, J.C.: Effect of martensite volume fraction on the mechanical behavior of an UNS S41003 dual-phase stainless steel. Mater. Sci. Eng. A-Struct. Mater. Prop. Microstruct. Process. **797**, 140208 (2020)
13. Alvarez, L.F., Garcia, C., Lopez, V.: Continuous cooling transformations in martensitic stainless steels. ISIJ Int. **30**, 516–521 (1994)
14. Guimarães, J.R.C., Rios, P.R.: Martensite starts temperature and the austenite grain-size. J. Mater. Sci. **45**, 1074–1077 (2010)
15. ASTM International: ASTM A240/A240M-11, Standard Specification for Chromium and Chromium-Nickel Stainless Steel Plate, Sheet, and Strip for Pressure Vessels and for General Applications. ASTM International, West Conshohocken (2019)
16. ASTM International: R 92-82, Test Method for Vickers Hardness of Metallic Material. ASTM International, West Conshohocken (2013)
17. ASTM International: E407 – 07, Standard Practice for Microetching Metals And Alloys. ASTM International, West Conshohocken (2015)
18. Back, J.G., Surreddi, K.B.: Microstructure analysis of martensitic low alloy carbon steel samples subjected to deformation dilatometry. Mater. Charact. **157**, 109926 (2019)
19. Verdiere, A., Cerda, F.M.C., Llanes, A.B., Wu, J.: Effect of the austenitizing parameters on the microstructure and mechanical properties of 75Cr1 tool steel. Mater. Sci. Eng. A **785**, 139331 (2020)
20. Shackelford, J.F.: Ciência dos materiais, 6th edn. Prentice Hall Brasil, São Paulo (2008)
21. Askeland, D.R., Phulé, P.P.: Ciência e engenharia dos materiais, 4th edn. Cengage Learning, São Paulo (2008)
22. Callister, W.D.: Ciência e Engenharia de Materiais: Uma Introdução, 5th edn. LTC, Rio de janeiro (2002)

Influence of Cutting Speed on Chip Formation and Hardening in the Deep-Hole Drilling Process of VTM-PLUS Tool Steel

Virgilio P. Ricci[1(✉)], Paulo H. S. Rodrigues[1], Gilberto L. A. Xavier[1], Amauri Hassui[2], and Carlos E. H. Ventura[1]

[1] Department of Mechanical Engineering, Federal University of São Carlos (UFSCar), Rod. Washington Luis Km 235, São Carlos, SP 13565-905, Brazil
virgilio@estudante.ufscar.br

[2] Department of Manufacturing and Materials Engineering, University of Campinas (UNICAMP), R. Mendeleyev 200, Campinas, SP 13083-970, Brazil

Abstract. This study aimed to evaluate the effect of v_c on the hardening of chips formed during the deep drilling process in VTM-PLUS tool steel, in the annealed state. The deep drilling was carried out with gun drills having straight channels, internal coolant, high speed steel shank, carbide tip and 7.8 mm diameter, with the ratio L/D = 26. The v_c of 60 and 70 m/min were analyzed. To evaluate the influence of cutting speed on the hardening of the chips they were analyzed via microscopy and Vickers hardness test. From the results obtained, it was observed that the formed chips were are segmented for both speeds and that the increase in the cutting speed caused an increase in its hardness. Hardness values increased in the condition of $v_c = 70$ m/min (321 ± 18 HV) when compared to the annealed condition (223 ± 3 HV). For the condition $v_c = 60$m/min there was also an increase in hardness values (309 ± 12 HV), although smaller, when compared to those of higher cutting speed. With the results obtained, it was possible to conclude that the increase in the deformation rate, imposed by the higher cutting speed, led to an increase in the hardness values due to the increase in work hardening.

Keywords: Deep hole drilling · cutting speed · chip formation

1 Introduction

Machining technology seeks to combine productivity, aiming at savings during the process, and higher quality in the finished product [1]. Drilling, when compared to operations such as turning and milling, is considered complex and becomes more complex with increasing hole length. The process is called deep hole drilling when the ratio between the length (L) and the diameter (D) of the hole is at least 10 times (L/D > 10). Deep drilling processes can produce holes with very low Rz surface roughness [1, 2].

Even for the best and most sophisticated machine tools, drilling a deep hole with a smooth surface texture, perfectly straight and with the required diameters from edge to edge is no easy task. In deep hole drilling processes, as the tool has a long length,

D. de Oliveira et al. (Eds.): COBEF 2023, LNME, pp. 95–104, 2024.
https://doi.org/10.1007/978-3-031-43555-3_9

it is common for its rigidity to be impaired and, thus, generate vibration, which can produce deviations in the dimensions of the machined hole, affect its surface roughness and roundness, causing, in addition, the tool breakage [3–5]. For a high-quality hole, different machining conditions must be defined, such as cutting speed, feed rate and type of cutting fluid [1, 6].

Chip formation in drilling can be influenced by some conditions, such as variable cutting speed along the drill radius, deformation, and consequent crushing of the material in the transverse edge region, modification of the wedge angle and the rake angle as a function of the radius, difficulty chip evacuation through channels [7–9]. During the drilling of ductile materials with small chip thickness, there is the formation of the chip in the shape of a conical spiral, while for large chip thicknesses, there are conical spiral chips with a large pitch. According to [8], the large pitch conical spiral chips occur when the chips lose the ability to maintain their compact shape, thus increasing the spiral pitch until the contact and friction with the walls of the drill channel restrict their displacement, generating its final form.

During the machining process, the material is subjected to high rates of deformation in the plastic region. These deformations produce some important effects in the microstructure of the steels, such as grain refining, which increases the number of contours per unit volume, and an increase in the number of punctate defects, making the movement of dislocations difficult and, consequently, the mechanical properties such as tensile strength and hardness are increased. Another factor that increases the hardening effect of steels is the presence of dispersed carbides in the matrix, which hinder the movement of dislocations. It should be noted that tool steels used for hot work have a large amount of carbides in their matrix [10].

Steels for hot work applications are subdivided into three classes, according to the type of alloy used to impart high hot hardness: (i) chromium steels for hot work, which nominally contain 5% Cr and significant amounts of other elements, including silicon, molybdenum and vanadium; (ii) tungsten steels for hot work and (iii) molybdenum steels for hot work [11, 12]. Chromium tool steels for hot work must have excellent resistance to impact at high temperatures, to softening during exposure to high temperatures and resistance to thermal fatigue, as they are constantly exposed to temperature variations. Medium-alloyed hot work tool steels such as H13 meet the compromise of both extremes: good hot-breaking resistance and reasonable toughness. This class of steels is quite suitable to produce stamping and forging dies, with which productivity is quite high [11].

After casting, tool steels are thermally treated to reach the necessary hardness and toughness, according to the required application. To carry out the drilling process efficiently, it is important that these steels have low hardness, being more common their use in the annealed state [11], with microstructure of spheroidized carbides uniformly dispersed in a ferrite matrix. This microstructure has low hardness, which facilitates machining. Annealing eliminates the martensite or pearlite microstructure that may have formed during cooling after hot working and homogenizes non-uniform deformation effects that may have developed during hot working of complex or heavy sections [11].

2 Methodology

The deep drilling tests were carried out in a Hermle Machining Center model C400 (Fig. 1), with maximum spindle speed of 18,000 rpm, maximum power of 20 kW and cutting fluid pressure of 80 bar, available in TecnoHow company facilities.

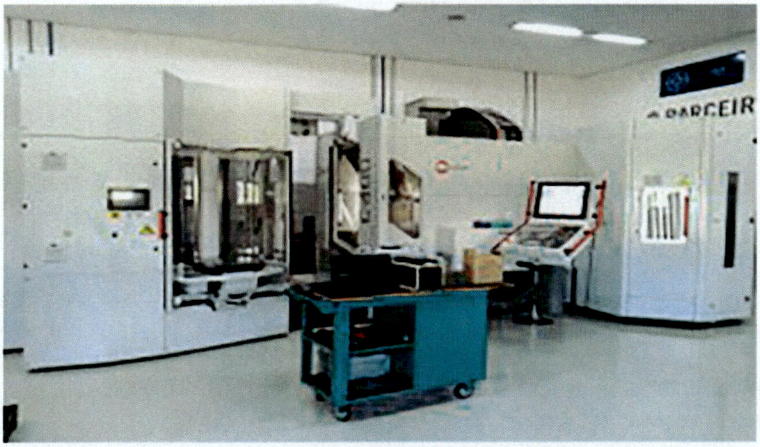

Fig. 1. Hermle C400 machining center used for the deep hole drilling tests.

The material used in this work was a block of VTM-PLUS tool steel in the annealed state, with dimensions of 202 mm × 202 mm × 270 mm, supplied by Villares Metals. This steel corresponds to an alternative to H13 steel and, therefore, has similar characteristics.

In Fig. 2 it is possible to see the tools used in the deep drilling tests. Figure 2a shows the gun drill with straight flutes, internal coolant, high speed steel shank, carbide tip and 7.8 mm diameter, with ratio L/D = 26. Figure 2b shows the drill tip view. When using gun drills, it is necessary to make a pre-hole (guide hole), with a depth approximately equal to its diameter, to facilitate its entry and prevent breakage. In Fig. 2c, the cutter used to make the pre-holes is shown, which had a diameter of 6 mm and 3 teeth. As the holes had a diameter of 7.8 mm and the milling cutter 6 mm, a circular interpolation was performed to reach a diameter of 7.8 mm. The tools were provided by ISCAR.

The through-holes were made continuously. As lubricating fluid was used an emulsion of vegetable oil Vasco 7000, supplied by Blaser, applied with a pressure of 80 bar. In the tests, the cutting speed was varied in two levels (60 and 70 m/min) and a feed rate of 0.02 mm/rev was kept constant.

For the microstructural characterization of the VTM-PLUS steel and the chips, these were mounted in bakelite, with subsequent finishing with 220, 400, 600 and 1200 mesh sandpaper followed by abrasive polishing with 1.0 μm alumina. Chemical attack was performed with 2% Nital reagent. The microstructural characterization was carried out with the aid of a Philips scanning electron microscope (SEM), model XL30 FEG, coupled with a semiquantitative analyzer by energy dispersive spectroscopy (EDS), available

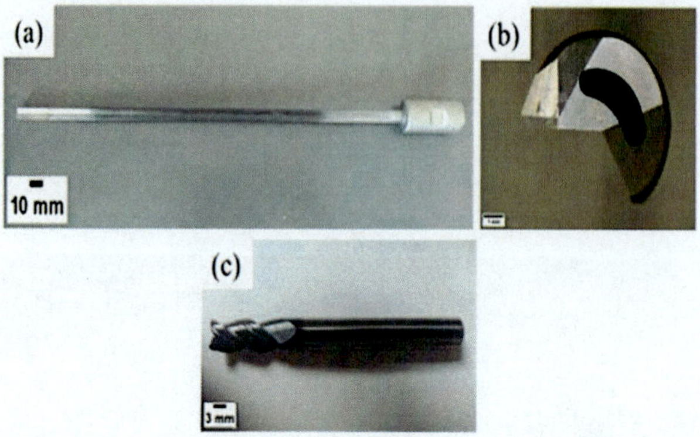

Fig. 2. Gun drill used in the experiments, (b) view of the drill tip and (c) milling cutter used for pre-holes.

at the Laboratory of Structural Characterization (LCE) at UFSCar. Figure 3 shows, illustratively, the method used to measure the chip shear angles. In the step of measuring the shear angle, an Olympus stereoscope, model SZ61, available at the Laboratory of Microstructure, Processing and Solidification Processes (M2PS), was used, in addition to the ImageJ software.

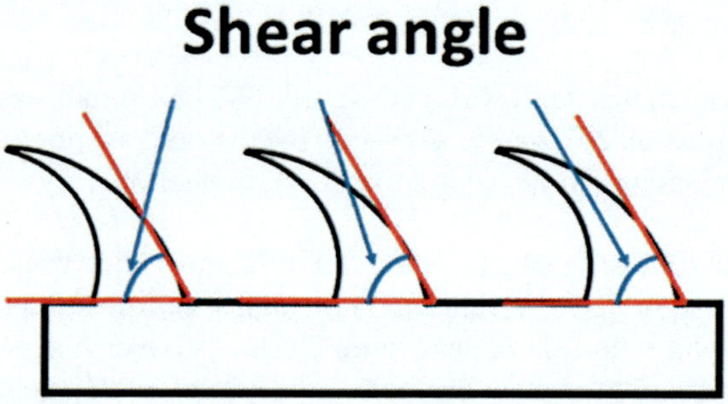

Fig. 3. Illustrative image of the method used to determine the shear angle of the chips.

The mechanical characterization, both of the VTM-PLUS steel in the annealed state, and of the chips obtained at the two cutting speeds tested, was performed using the Vickers microhardness test, in a Shimadzu hardness tester, model HMVG20ST, available at the Laboratory of Microstructure, Processing and Solidification Processes (M2PS), with a load of 0.500 kgf applied for 15 s using a 40x magnification lens. To assess the hardness measurements, 15 measurements were performed in each condition.

3 Results and Discussions

As a starting point, it is possible to verify the microstructure of the VTM-PLUS steel used in this work in the annealed state (as received), shown in Figs. 4a and 4b. In this state, the material presents a ferritic matrix (dark part) of low hardness, in which a considerable amount of spheroidized carbides (small, light-colored particles), fine and homogeneously dispersed, is present.

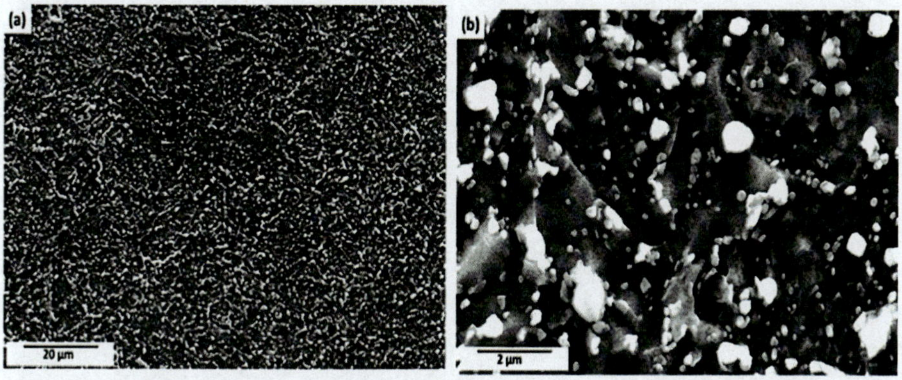

Fig. 4. Microstructure of VTM-PLUS steel in the annealed state.

In Table 1 it is presented the results referring to the semiquantitative analysis of energy dispersive spectroscopy (EDS), performed on VTM-PLUS steel. With these results, it is obvious that the element in greater quantity is Iron (Fe), with 89.2 wt%. On the other hand, Vanadium (V) is the element in the lowest concentration, with 0.8% by mass.

Table 1. Results of the semi-quantitative EDS analysis of the VTM-PLUS steel.

Elements	Wt %
Iron (Fe)	89.2
Carbon (C)	0.7
Chromium (Cr)	4.9
Molybdenum (Mo)	2.5
Vanadium (V)	0.8
Tungsten (W)	1.9

In Figs. 5a–f it is presented the mapping of the elements present in the VTM-PLUS steel, as shown in Table 1.

The results obtained in this work are similar to those observed by KANG et al., (2015), who used an H13 type tool steel. It was also possible to note the presence of dispersed carbides in the ferritic matrix, making it possible to verify that these carbides are of the M2C and/or M6C type, that is, they are rich in Molybdenum (Mo) (Fig. 5d). In the region where the carbides are found, there is a greater presence of Mo and an impoverishment of Fe.

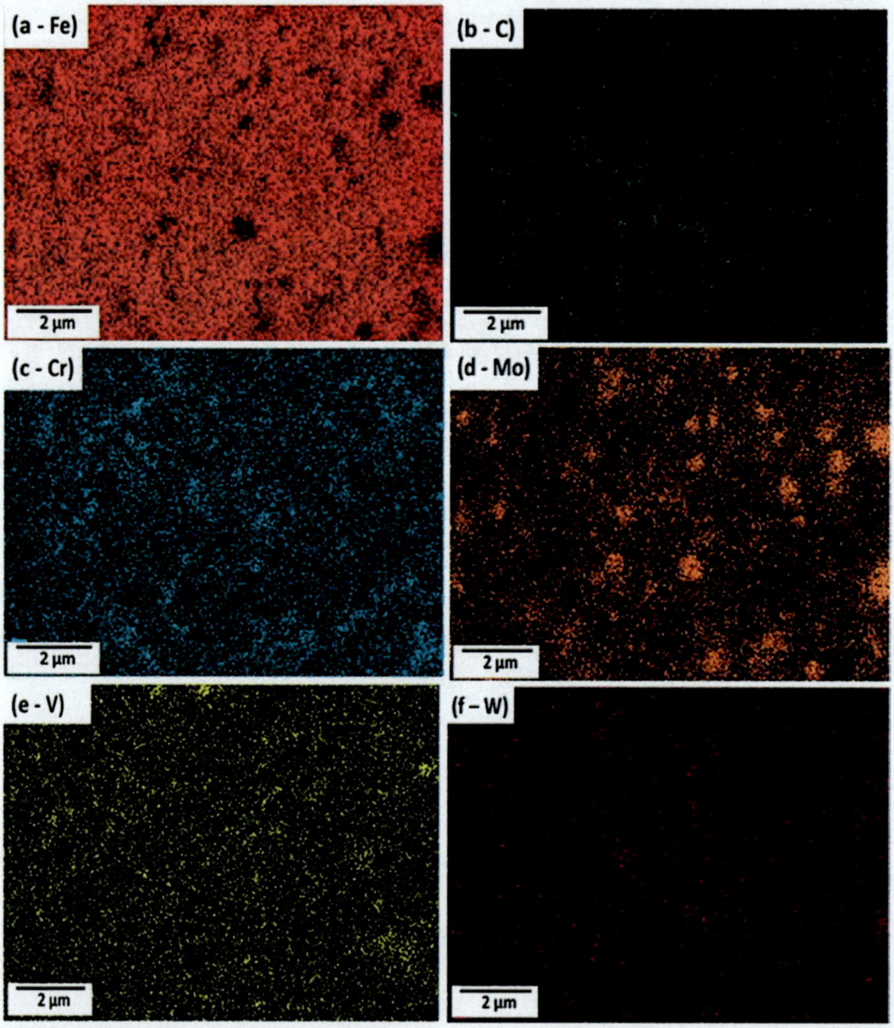

Fig. 5. Mapping of the distribution of elements present in VTM-PLUS steel.

In Figs.6a and 6b it is possible to see the micrographs of the chips generated with cutting speeds of 60 and 70 m/min, respectively, after the deep hole drilling process. As seen in Figs. 3a and 3b, the spheroidized carbides were present (light spots) dispersed in the ferritic matrix (dark part). Since the machining parameters were light and the workpiece material was annealed, one possible hypothesis is that the process was carried out cold, with a temperature lower than half of its melting temperature (T < 0.5 × Tf) and, therefore, the presence of martensite induced by the process was not observed.

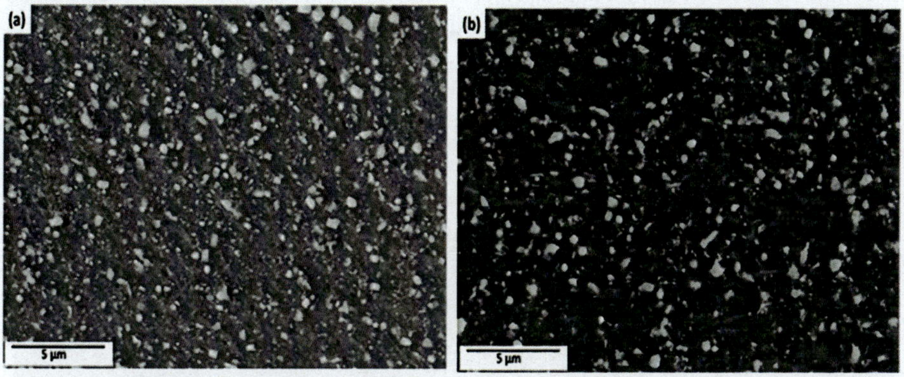

Fig. 6. Microstructure of chips formed with (a) $v_C = 60$ m/min and (b) $v_C = 70$ m/min.

The results obtained are in line with the literature [13] since, at higher speed, the strain rate imposed on the material grows and, consequently, the same happens to the density of dislocations, which already have their movement hampered by carbides dispersed in the matrix [10]. Considering that deep drilling is a cold process, the effect of dynamic recrystallization does not occur [14] and the hardening effect due to the movement of dislocations becomes more pronounced, due to their accumulation in spheroidized carbides [10] (Fig. 7).

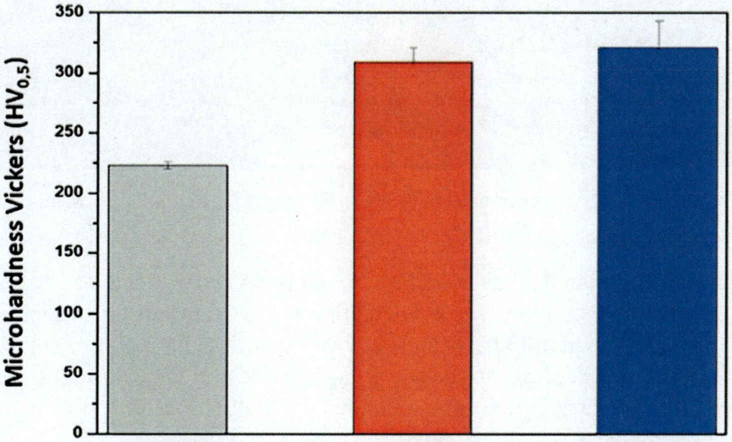

Fig. 7. Mean Vickers microhardness values (HV0.5) of annealed VTM-PLUS steel (grey) and chips obtained with $v_C = 60$ m/min (red) and 70 m/min (blue).

One way to analyze the influence of the deformation rate imposed on the material during machining consists in evaluating the shear angle of the formed chips. Initially, it was observed that these chips were quite segmented, as shown in Figs. 8a and 8b for the condition $v_c = 60$ m/min and in Figs. 8c and 8d for the condition $v_c = 70$ m/min. This is an indication that greater material deformation occurs with the application of higher cutting speeds, due to the higher strain rate imposed. Still looking at Figs. 8b and 8d, it can be seen in the higher speed condition the presence of a greater amount of deformation bands compared to the lower speed condition, which is another indication of the lower strain imposed in the latter case.

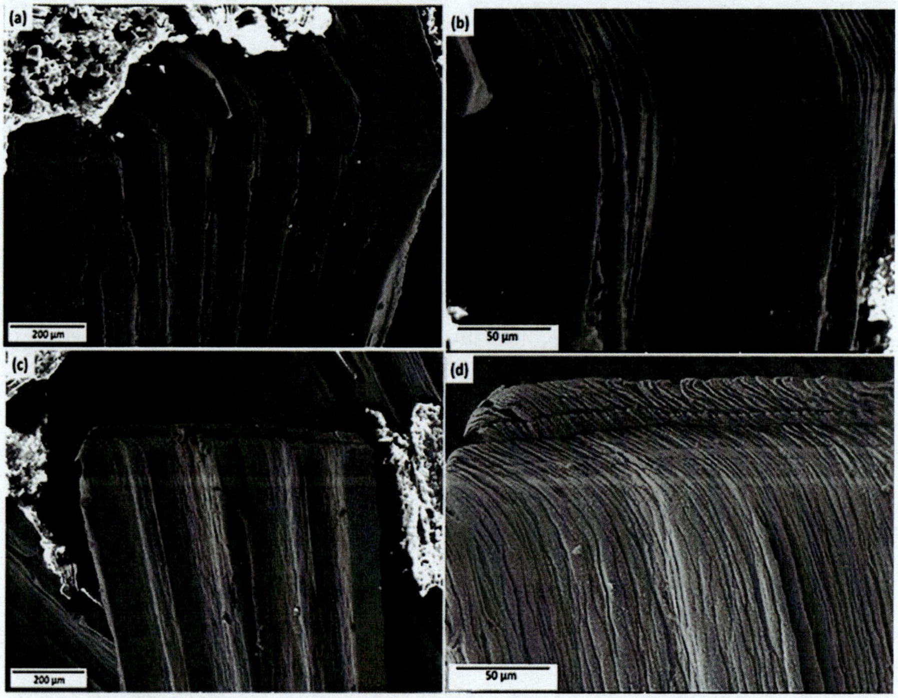

Fig. 8. Chip morphology in the condition of Cs = 60 m/min (a, b) and Cs = 70 m/min (c, d).

The chips produced in the first condition ($v_c = 60$ m/min) presented a shear angle of 37.4° ± 3.3°, greater than in the second condition ($v_c = 70$ m/min), in which an angle of 27.1° ± 3.1°, as shown in Figs. 9a and 9b where it is possible to observe the general view of the chips. The decrease in the shear angle is another indication that the higher cutting speed induces greater deformation in the VTM-PLUS steel.

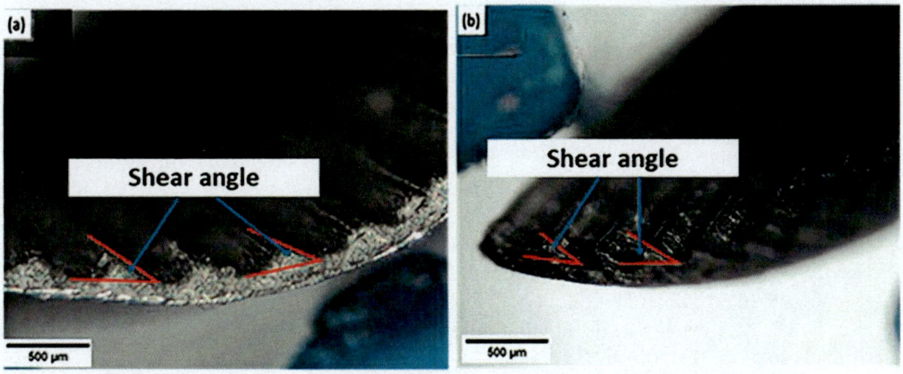

Fig. 9. General view of the chips (a) $v_C = 60$ m/min; (b) $v_C = 70$ m/min.

4 Conclusions

Through the tests carried out in this work, it was possible to conclude that:

- The VTM-PLUS steel in the annealed state, microstructurally, is formed by a ferritic matrix with globular carbides, mainly molybdenum, finely and homogeneously dispersed.
- Increasing the cutting speed caused a higher deformation rate which, in turn, increased the hardness values, although it did not cause a visible change in the material's microstructure.
- In the condition of higher cutting speed (70 m/min), the shear angle was smaller and the occurrence of a greater amount of shear bands in the chip was verified, which indicates a greater degree of deformation.

Acknowledgments. The authors would like to thank FUNDEP (Foundation for Research Development) for financial support to the project "Nationalization of tool steel dies for hot stamping parts for the automotive segment", through the ROTA 2030 Program – Line IV. The authors also thank the companies Blaser, Iscar, TecnoHow and Villares Metals for providing the materials and infrastructure to carry out the experiments.

References

1. Aized, T., Amjad, M.: Quality improvement of deep-hole drilling process of AISI D2. Int. J. Adv. Manuf. Technol. **69**(9–12), 2493–2503 (2013)
2. Liu, Z., Liu, Y., Han, X., Zheng, W.: Study on super-long deep-hole drilling of titanium alloy. J. Appl. Biomater. Funct. Mater. **16**(1), 150–156 (2018)
3. Chaudhari, A., Malarvizhi, S., Woon, K.S., Senthil Kumar, A., Rahman, M.: The effects of pilot hole geometry on tool-work engagement efficacy in deep hole drilling. J. Manuf. Process. **19**, 135–141 (2015)
4. Kong, L., Cao, S., Chin, J.H., Si, Y., Miao, F., Li, Y.: Vibration suppression of drilling tool system during deep-hole drilling process using independence mode space control. Int. J. Mach. Tools Manuf **151**, 103525 (2020)

5. Si, Y., Zhang, Z., Kong, L., Zheng, J.: Condition monitoring of deep-hole drilling process based on improved empirical wavelet de-noising and high multiple frequency components of rotation frequency. Int. J. Adv. Manuf. Technol. **114**(7–8), 2201–2214 (2021)
6. Baghlani, V., Mehbudi, P., Akbari, J., Nezhad, E.Z., Sarhan, A.A.D., Hamouda, A.M.S.: An optimization technique on ultrasonic and cutting parameters for drilling and deep drilling of nickel-based high-strength Inconel 738LC superalloy with deeper and higher hole quality. Int. J. Adv. Manuf. Technol. **82**(5–8), 877–888 (2016)
7. Han, C., Zhang, D., Luo, M., Wu, B.: Chip evacuation force modelling for deep hole drilling with twist drills. Int. J. Adv. Manuf. Technol. **98**(9–12), 3091–3103 (2018)
8. Ke, F., Ni, J., Stephenson, D.A.: Continuous chip formation in drilling. Int. J. Mach. Tools Manuf. **45**(15), 1652–1658 (2005)
9. Ke, F., Ni, J., Stephenson, D.A.: Chip thickening in deep-hole drilling. Int. J. Mach. Tools Manuf. **46**(12–13), 1500–1507 (2006)
10. Chen, R., et al.: The carbides, tensile properties, and work-hardening behavior of annealed H13 die steels with varied yttrium contents. Mater. Sci. Eng. A **806**, 140856 (2019)
11. Kang, M., Park, G., Jung, J.G., Kim, B.H., Lee, Y.K.: The effects of annealing temperature and cooling rate on carbide precipitation behavior in H13 hot-work tool steel. J. Alloy. Compd. **627**, 359–366 (2015)
12. Nanesa, H.G., Boulgakoff, J., Jahazi, M.: Influence of prior cold deformation on microstructure evolution of AISI D2 tool steel after hardening heat treatment. J. Manuf. Process. **22**, 115–119 (2016)
13. Sun, J., Ke, Q., Chen, W.: Material instability under localized severe plastic deformation during high speed turning of titanium alloy Ti-6.5AL-2Zr-1Mo-1V. J. Mater. Process. Technol. **264**, 119–128 (2019)
14. Mondelin, A., Valiorgue, F., Rech, J., Coret, M., Feulvarch, E.: Modeling of surface dynamic recrystallisation during the finish turning of the 15–5PH steel. Procedia CIRP **8**, 311–315 (2013)

Design, Implementation and Evaluation of a Wire Feeding System with Induction Heating Applied to L-DED

Manoel Kolling Dutra, Milton Pereira(✉), Fillipe Augusto Felipim Lavaqui, and Tiago Colvara Faria

Precision Engineering Laboratory – Laser Division (LMP-Laser), Department of Mechanical Engineering, Universidade Federal de Santa Catarina, Florianópolis, Santa Catarina 88040-900, Brazil

manoel.kolling@posgrad.ufsc.br, milton.pereira@ufsc.br, lavaquifellipe@gmail.com, tiagocf1208@gmail.com

Abstract. Laser Energy Directed Deposition (L-DED) process is a well-established industrial technique for precise coating deposition and additive manufacturing. Laser cladding aims to improve surface performance, thereby extending the lifespan of many components in severe corrosive and abrasive wear environments. Wire laser cladding consists of feeding the deposited material with a wire, which is fused by a laser energy source and manipulated over the deposition volume. This process offers several advantages over powder-based systems, such as a cleaner processing environment, reduced environmental cost for wire production, no waste of additional material, and a higher material deposition rate. Through innovative equipment, such as dedicated wire feeding systems and preheating techniques, customized manufacturing strategies can improve coating quality and productivity. Among recent innovations, preheating the wire results in lower energy rates delivered by the laser. This type of heating is called hot-wire heating. Hot-wire is well-known in arc welding, where its synchronization with the arc's electrical parameters provides a dramatic increase in metal transfer stability and controllability. Based on this, the present paper addresses the development of a wire feeder with integrated induction heating. A geometric characterization of the deposited strands and a characterization of the wire heating temperature profile with thermographic imaging measurement were performed to validate the developed device and its peripherals regarding deposition stability.

Keywords: L-DED · coating · hot-wire · induction heating

1 Introduction

Laser materials processing has gained increasing prominence in various industrial applications such as welding, coating, and additive manufacturing, among others, due to its relatively low heat input, precise heating, low distortion, and high processing flexibility [1]. In surface coating applications, an additional material is inserted into the process to generate deposited layers, which can be deposited from wire or powder.

D. de Oliveira et al. (Eds.): COBEF 2023, LNME, pp. 105–113, 2024.
https://doi.org/10.1007/978-3-031-43555-3_10

Wire laser cladding involves feeding the deposited material with a wire that is fused by a laser energy source and manipulated over the deposition volume [2]. This process offers several advantages over powder-based systems, such as a cleaner processing environment, reduced environmental cost for wire production, no addition in material waste, and a higher material deposition rate [3]. However, wire deposition poses some challenges due to the precise requirements for feeding the wire into the laser focus point, ensuring the melting of the wire and base metal simultaneously, and generating uniform and controlled deposition [4].

One technique that is gaining momentum to improve process stability is wire preheating or hot-wire [5]. Preheating the wire reduces the relative positioning requirements between the wire and the laser beam. Additionally, it achieves higher deposition efficiency and reduces heating damage to the base metal during the process [6].

Therefore, the development of equipment that can perform wire deposition in a more stable manner and integrate techniques such as hot-wire has been gaining ground in the literature to improve and optimize wire deposition processes [7]. This paper aims to contribute to this area by presenting the development of a wire feeding system capable of preheating the wire through an inductive circuit for use in L-DED.

2 Methodology

The system design was developed based on the adapted PRODIP methodology [8]. During the informational design phase, stakeholders, their needs, and project requirements were defined, as well as the importance order for each requirement. Four mechanical design requirements were established to ensure the satisfactory performance of the wire feeder:

1. The clamping base structure must be rigid enough to exhibit negligible deformation and vibration during operation.
2. The alignment of the wire in relation to the laser beam focus must ensure stability during the deposition process.
3. The coil module and electronic circuit must be as close as possible to the working area to perform induction preheating of the deposited wire.
4. The wire and powder heads must be interchangeable for use on the machine.

Regarding the electro-electronic requirements, the following points were considered:

1. The wire feeder must have an inductive heating system.
2. It must have hardware capable of feeding the wire.
3. It must be able to supply the necessary electrical power to drive the induction coils.

Subsequently, the project was divided into two fronts: the mechanical design and the electro-electronic design. The concepts and their analyses were carried out simultaneously with development to provide greater speed and dynamism to the project.

2.1 Mechanical Design

In this step, the initial goal was to ensure the structural rigidity of the system. Therefore, a fixture was designed consisting of a base made of square cross-section tubes measuring

30 mm x 30 mm with a wall thickness of 2 mm. The tubes were joined together using the MIG/MAG welding process to form an arm that ensures parallelism between the device and the laser head.

A 3 mm steel plate was attached to this primary structure to enable the assembly and positioning of the other systems of the wire feeder, such as the traction and steering system, and coils. Finally, the entire assembly is fixed to the laser head using the primary structure in a manner that allows for the system to be removed for maintenance and use of the powder head in another configuration of the system.

To move the wire, a servo motor is utilized with an elastic coupling on its shaft and a spacer that allows for the correct positioning of the assembly on the metal plate. On the opposite side of the plate, the wire feeder is mounted, so that its shaft is screwed into the elastic coupling of the motor. Figure 1 shows the proposed model of the mounting structure and coupling system between the motor and the wire feeder.

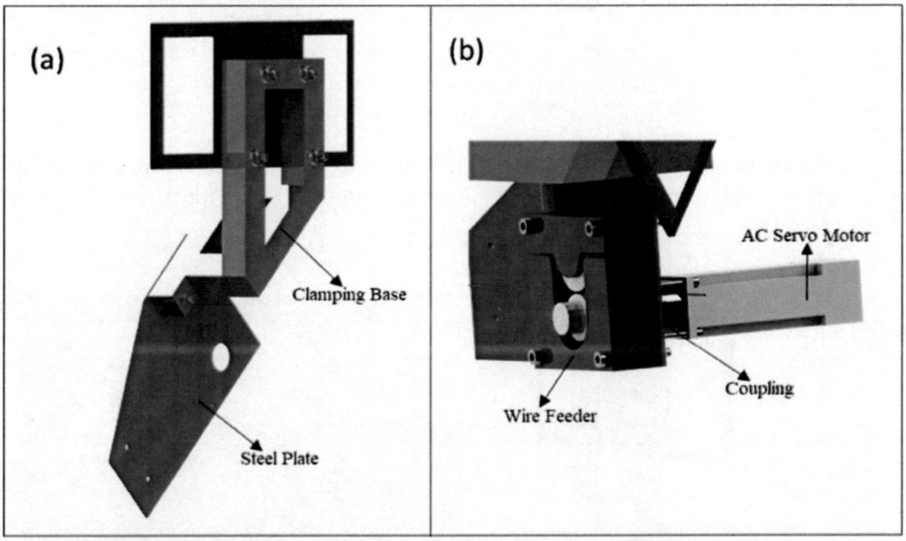

Fig. 1. Model of the mechanical structure (a) and the coupling system (b).

One of the most crucial aspects related to the system's construction was to ensure that the wire remained aligned with the laser beam during processing. To meet this requirement, two contact nozzles tips of welding equipment were used after the wire feeder, with an opening equal to the wire diameter of 1.2 mm, to align the wire and keep it in the same plane as the laser beam.

To refine the positioning of the feeder and allow the wire to work precisely in the desired position aligned with the laser beam, a positioner with three degrees of freedom and fine adjustment, based on worm screws, was implemented.

To make the assembly compact and fit the heating coil into the limited space imposed by the wire steerer, the heating coil was positioned in that area. As the proposed heating mode is accomplished by electromagnetic induction, it is crucial that there is no field conductor inside the coil, except the wire itself.

Figure 2 illustrates the device to ensure the alignment of the wire with the beam in (a) and the position of the induction heating coil in (b).

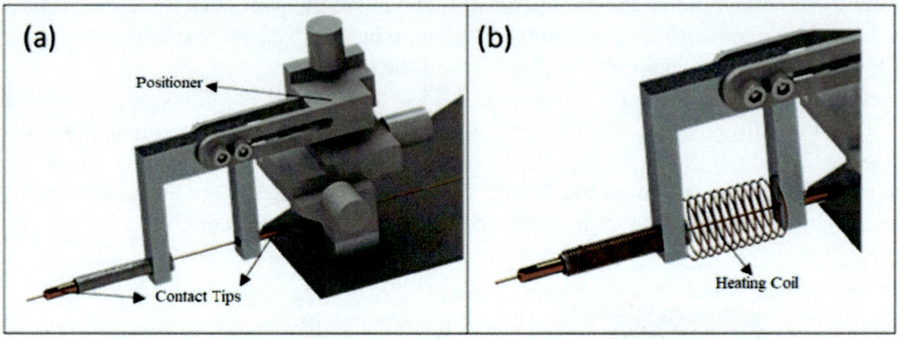

Fig. 2. Position adjustment system (a) position of the inductive heating coil (b).

The concepts and solutions developed fulfilled the specified design requirements, resulting in the final design, which was manufactured and integrated into the laser system for the prototype assembly, as depicted in Fig. 3.

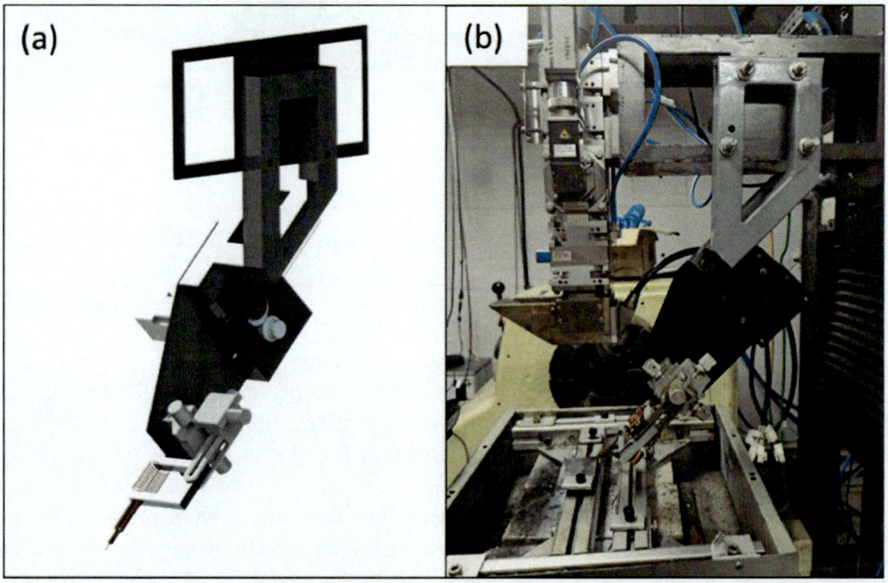

Fig. 3. Designed (a) and built (b) wire feeding system.

2.2 Electro-electronic Design

To design the electrical project, the electro-electronic assembly was divided into three parts:

1. Servomotor drive for wire traction
2. Inductive preheating system
3. Mechatronic integration

For wire displacement, a PANASONIC servo motor model CA MSMD012P1T was chosen due to its 10,000 steps per revolution resolution and speed of up to 3000 rpm. To drive the servo motor, a MINAS A4 control module with integrated PID control was used. This module is powered by a continuous source and receives pulses from an Arduino, which commands the speed and direction of the motor's rotation and transforms this information into a 220V AC power signal that moves the motor shaft.

For the pre-heating circuit, an LC circuit was designed, consisting of a parallel capacitor and inductor with two MOSFETs connected at their ends to switch the ground reference signal of the circuit, maintaining the current in the system in an oscillatory dynamic. The heating coil has a central tap that receives a positive DC voltage. This voltage charges the capacitor, causing the first MOSFET to stop the current flow, and the second MOSFET to enter the conduction regime, inverting the reference signal. This reversal causes the inductor to induce a reverse voltage to oppose the change, charging the capacitor with its polarity reversed, and causing the MOSFETs to switch again. This cycle repeats continuously, resulting in a rapid variation of the field in the inductor, operating around a 250 kHz frequency.

The magnetic field variation in the inductor causes inductive heating of the wire through the eddy currents induced in the wire. The high frequency imposed in the LC oscillator circuit is necessary due to the wire diameter and edge effect. Magnetic saturation of the material occurs, and the total heating efficiency decreases if not respected.

Figure 4 (a) shows the preheating circuit with its source and the circuit for controlling and feeding the servomotor. Figure 4 (b) displays the preheating coil with the wire passing through it during the process.

3 Results and Discussion

To evaluate the mechanical design and control hardware for wire feeding, a built prototype underwent a series of tests. The first test aimed to verify the repeatability of the process by depositing nine strands on SAE 1020 carbon steel plates using ER70S-06 wire, and measuring their lengths. This material was chosen due to its compromise between magnetic characteristics and its application in deposition processes. The parameters used were: 1300 W laser power, 1.2 mm spot size with Z + 35 mm blurred laser, 300 mm/s laser scan speed, and an average wire speed of 0.8 m/min (which falls within the range of wire speeds typically used, 0.5 to 1.5 m/min) [5, 7]. The CNC control of the machine was programmed to produce a cord length of 30 mm, which results in an expected final length of 31.2 mm for the strand. This is because the laser beam is positioned in the center of the wire, and half the wire diameter must be added to each side

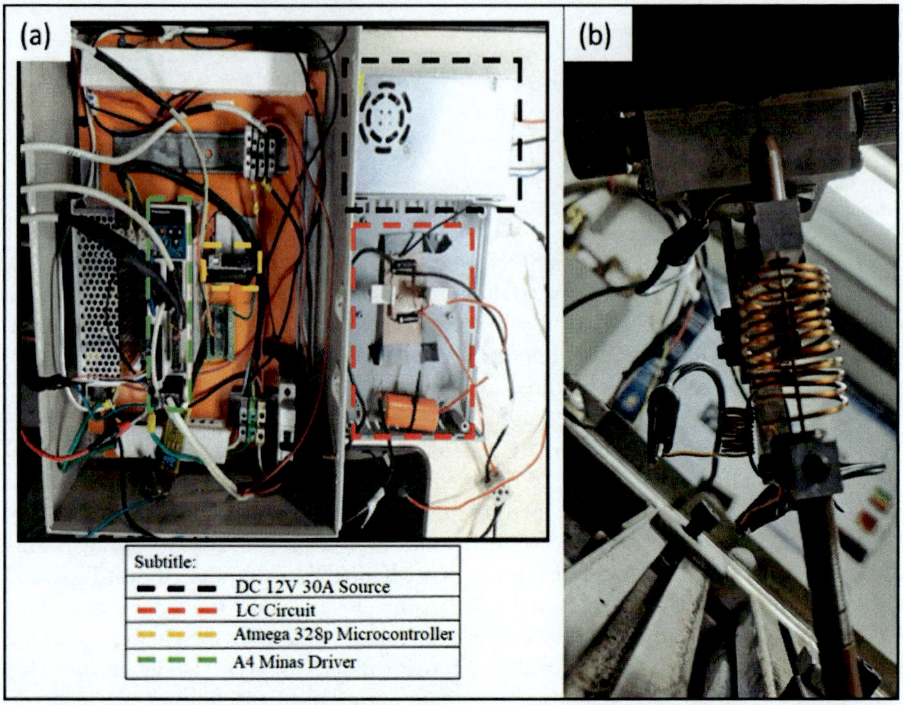

Fig. 4. Electro/electronic circuit (a) and preheating coil (b).

of the strand. The system proved to be robust, resulting in a bead size of (31.15 ± 0.35) mm. Figure 5 shows (a) the deposited strands that were subjected to measurements, and (b) a graph comparing the average value of the strand length with the expected value, along with representation of the error in the error bar plotted on the graph.

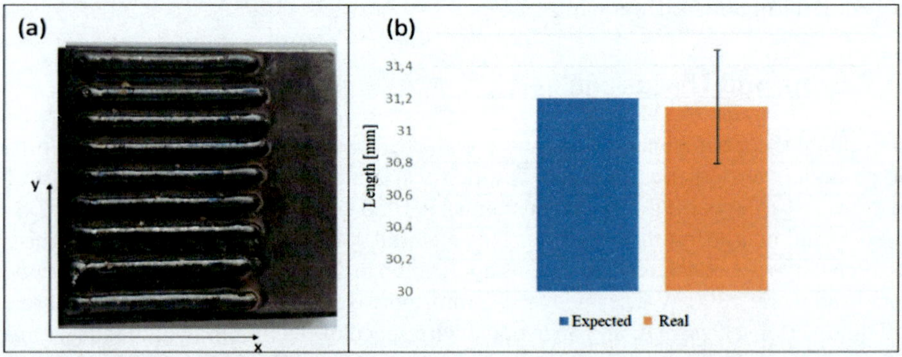

Fig. 5. Deposited strands (a) and measurement results (b).

Another factor that was analyzed was the ability of the mechanical and electronic wire feed system to perform both stranding and coating, as coating is the most common practice in the intended applications for the system. To evaluate this capability, a coating experiment was performed with 10 beads having a 30% overlap between them. Figure 6 shows the results of this experiment, where a relatively stable process was observed, although some discontinuities between the strands were still present. These discontinuities can be mitigated with a better study of the process parameters, specifically adjustments in the laser emission power and deposition speed.

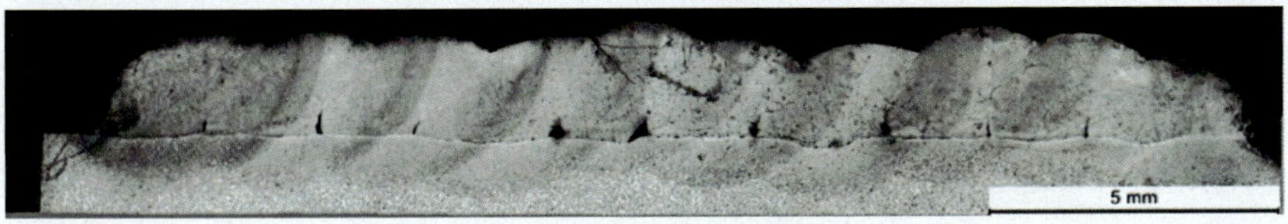

Fig. 6. Coating Macrography.

In addition to the material deposition validations mentioned earlier, the proposed device was tested to verify its wire movement control capability. The device was tested under no load for dynamic feeding conditions, with the aid of a resistive encoder attached to the feeding shaft to monitor the quality of the wire displacement response. It was observed that the system was able to move in both directions smoothly, with no abrupt variations in feed speed during both the wire feed and recoil periods, thus providing a stable dynamic feed control process for these boundary conditions. Figure 7 shows the monitored values of displacement versus time plotted on a graph, which validates the dynamics of wire movement.

After validating the rigidity of the wire feeder mounting bracket and the wire movement process, the induction heating circuit was evaluated for its ability to heat the wire. This evaluation was done in two ways: through thermographic and indirect measurement using K-type thermocouples. For the thermocouple reading, the heating system was already in permanent regime, and the induction coil responsible for pre-heating was activated with the following parameters: 12 A of electric current at a frequency of 250 kHz. After 2 min, the temperature was measured by touching the thermocouple to the heated wire, resulting in a temperature of (473 ± 2) ºC. In the measurement using a thermal camera, the heating was filmed for 2 min and temperatures were obtained at 3 positions in the system: in the middle of the coil, at the end of the coil, and in the surrounding environment. The thermal images were analyzed, and a temperature profile graph was generated, as shown in Fig. 8.

The temperature values measured and presented by the induction heating system are capable of heating the wire to acceptable values for the hot-wire process. In the research conducted by Voigt, Cunha, and Niño (2020) [7], the authors proposed using induction heating for wire preheating in the TIG process, which uses electric arc. They used a commercial current source from the manufacturer IMC dedicated for induction heating, and the temperature values ranged from 100 to 700 °C. In the work of Huang *et al.* (2022) [5], the authors achieved temperatures up to 800 °C in wire preheating using

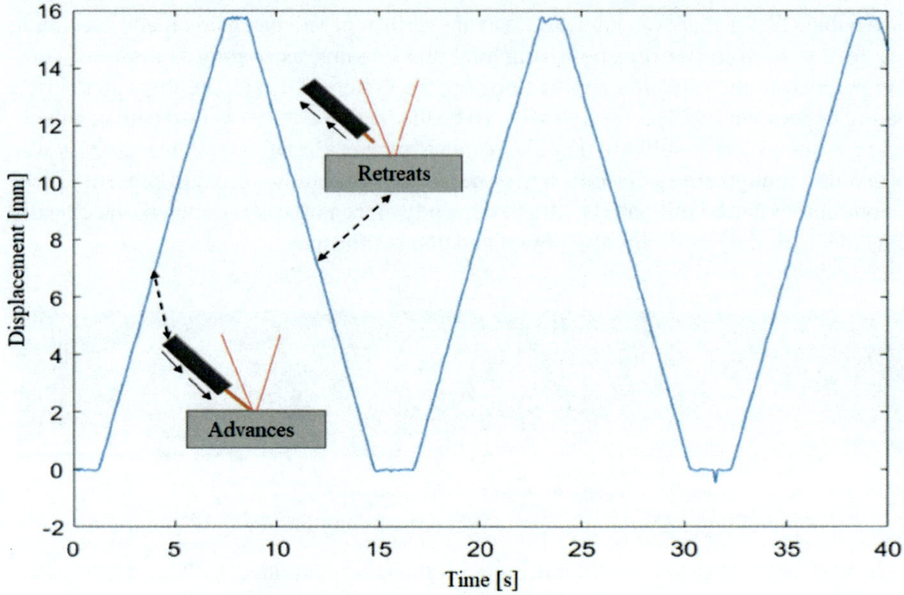

Fig. 7. Validation of the wire feeding and handling system.

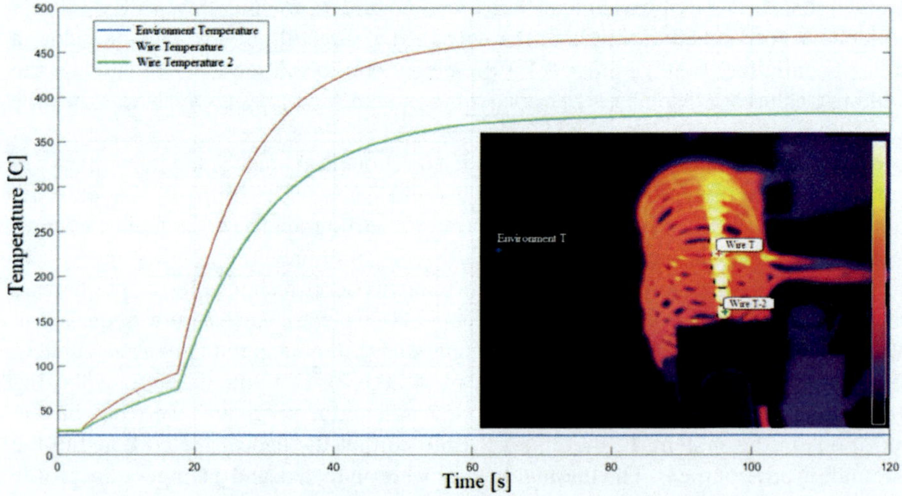

Fig. 8. Validation of induction heating system.

a controlled current source from the company Fronius and resistive heating instead of inductive heating. The aforementioned works prove that the inductive preheating device proposed in this study is relevant for its heating capability and application in L-DED processes.

4 Conclusion

The wire feeder with integrated induction heating proposed in this work was able to perform depositions as per the tests performed, indicating that the system meets the necessary technical specifications for the proposed depositions. Additionally, it offers the possibility of performing coatings with satisfactory visual and geometric aspects. However, it is recommended to optimize the deposition parameters and analyze the generation of discontinuities, the adhesion, and the chemical resistance of the deposits.

The induction heating system has demonstrated its ability to reach common hot-wire temperatures ranging from 100 to 800 °C, as shown in the cited works. Furthermore, the integrated aspect of wire feeding and preheating in one device makes it highly applicable in L-DED deposition processes.

The system is also designed for easy interchangeability, as changing between the powder and wire heads requires the removal of only four screws. This feature allows for convenient switching between systems and greater flexibility for the operator.

Acknowledgments. We would like to express our gratitude to CNPq for the support to this work.

References

1. Poprawe, R.: Tailored Light 2: Laser Application Technology, 1st edn. Springer, Berlin, Heidelberg, Germany (2011)
2. Kisielewicz, A., et al.: Hot-Wire Laser-Directed Energy Deposition: Process Characteristics and Benefits of Resistive Pre-Heating of the Feedstock Wire. Metals **11**(4), 634 (2021)
3. Segerstark, A.: Laser Metal Deposition using Alloy 718 Powder Influence of Process. Materials Characterization (Vol. 63), (2017)
4. Peng, W., Jiguo, S., Shiqing, Z., Gang, W.: Control of wire transfer behaviors in hot wire laser welding. Int. J. Adv. Manuf. Technol. **83**(9–12), 2091–2100 (2016)
5. Huang, W., Xiao, J., Chen, S., Jiang, X.: Control of wire melting behavior during laser hot wire deposition of aluminum alloy. Opt. Laser Technol. **150**(2), 107978 (2022)
6. Zhao, S., Xu, S., Huang, Y., e Yang, L.: Laser hot-wire cladding of Ni/WC composite coatings with a tubular cored wire. Journal of Materials Processing Technology, 298(4), 117273 (2021)
7. Voigt, A.L., da Cunha, T.V., Niño, C.E.: Conception, implementation and evaluation of induction wire heating system applied to hot wire GTAW (IHW-GTAW). J. Mater. Process. Technol. **281**(1), 116615 (2020)
8. Back, N., Ogliari, A., Dias, A., Silva, J.C.: Projeto integrado de produtos: planejamento, concepção e modelagem. (p. 629). Editora Manole Ltda. Avenida Ceci, 672 Tamboré, Barueri, Brasil (2010)

Geometry Study on 410NiMo Alloy Parts Printed by WAAM-CMT

Matheus André de Araújo Soares, André Luiz Brito Novelino, and Maksym Ziberov(✉)

Faculty of Technology, Department of Mechanical Engineering, University of Brasilia, Brasilia 70910-900, Brazil
mziberov@unb.br

Abstract. Within metallic additive manufacturing, the Wire + Arc Additive Manufacturing (WAAM) method is one of the processes that is of great interest to the industry, based on layer-by-layer deposition. This study aims to present the best deposition parameters of the 410NiMo stainless steel alloy, using the WAAM method, and more specifically the Cold Metal Transfer, taking as a comparison the analysis of the best geometry obtained. The desired geometry in an additive manufacturing process is the one with the largest effective area, that is, with the greatest use of material. For this, depositions of single beads and with 5, 15 and 30 layers were carried out, one continuously and the other deposited in two stages with an intermediate cooling time (15 + 15 layers), varying the parameters between them, followed by analysis of their height and width, in addition to measuring the wettability angle obtained, in order to obtain the part with better geometries. In the case of wall depositions, using the parameters obtained from the analysis of single beads, it was verified that the geometry of the samples varied according to the deposited layers, possibly caused by the low temperature of the substrate, causing the molten material to solidify faster. For samples with 30 layers, the depositions without interruption of the arc, with the deposition in stages were compared, the latter being the strategy that presented a greater height, with a narrower width, generating a better use of material and geometry.

Keywords: Additive Manufacturing · WAAM · CMT · 410NiMo

1 Introduction

As manufacturing technologies have advanced, several complex, or economically unfeasible, processes have become viable for application in several industries, motivated by factors such as process automation and the quality and precision of new equipment. Among these technologies, metal additive manufacturing has seen a significant expansion in the market, with predictions that the global market will reach $11.45 billion by 2030 [1]. This process also enables the manufacturing of complex geometries with a significant reduction in material waste compared to traditional processes and is compatible with automated applications proposed by Industry 4.0 [2].

D. de Oliveira et al. (Eds.): COBEF 2023, LNME, pp. 114–125, 2024.
https://doi.org/10.1007/978-3-031-43555-3_11

One of the technologies used for metal additive manufacturing is Wire + Arc Additive Manufacturing (WAAM), which utilizes concepts and equipment originally applied in welding, using a power source and wire feeding for thermal input and deposition [3]. Among the deposition technologies that can be used in the WAAM process, the most commonly used technology today is Cold Metal Transfer (CMT), a variant of the Gas Metal Arc Welding (GMAW) technology, developed by Fronius International GmbH. It includes a short-circuit deposition similar to GMAW, but with the use of a mechanical droplet cutting [4]. This process results in greater deposition stability, facilitating reproducibility and increasing energy efficiency, due to lower thermal input when compared to conventional deposition processes [5].

The WAAM process is applicable to several materials, with research conducted on commonly used alloys, ER70S-6 as an example [6], as well as other steel, aluminum, and titanium alloys. Among compatible materials, stainless steels have gained significant research focus, such as the martensitic stainless-steel ER410NiMo, which has high mechanical strength and corrosion resistance [7]. However, the weldability of such steels can be a challenge in the manufacturing process, requiring the determination of ideal deposition parameters to ensure proper material fusion with the deposited bead while maintaining an appropriate geometry of the manufactured part.

In order to explore the applicability and efficiency of the WAAM process for manufacturing stainless steel components using ER410NiMo, this study aims to investigate deposited walls using WAAM-CMT. The deposition quality will be evaluated based on the geometry obtained, by measuring the total and effective height and width of the samples. Temperature control will also be analyzed, focusing on the differences between each deposition strategy used for depositing the walls. Thus, the deposition parameters that result in the best geometry will be determined.

2 Materials and Methods

The deposition of the samples was carried out using a 1.2 mm diameter wire of AWS ER410NiMo. The CMT synergic method was used for the deposition, using a Fronius Transpulse Synergic 5000 CMT as power source. The power source parameters were regulated using a remote-control unit RCU 5000i from the same manufacturer. A Schneider Electric cartesian robot (model MAXR23-S42-C42) was operated with a G-code based program, depositing onto a carbon steel substrate with 6.35 mm thickness and 200 × 200 mm area. The shielding gas used during deposition was a mixture of Ar + 18%CO_2, with a flow rate of 15 l/min. After the deposition process, all samples were monitored using a KLX digital infrared laser thermometer with a resolution of 0.10 °C and accuracy of ±1.50 °C, adjusted for a blackbody emissivity. Also, all samples were cooled by natural convection.

The deposition plan consisted of initially fabricating seven single beads samples, each 50 mm in length, using a different deposition configuration to determine the best configurations to be used in the deposition of the walls. Once the initial parameters were defined, walls of 5, 10, and 15 layers with a length of 100 mm were deposited continuously (without arc interruption). In addition to the variation of current, welding speed, and wire feed speed, step-over distance between passes was also a parameter analyzed for the wall samples.

To produce 30-layer wall, three deposition methods were applied: continuous arc deposition, step deposition, with intermediate cooling to 100 °C at 15 layers, and deposition with interlayer temperature control in the range of 45 °C to 55 °C. These different methods were compared by evaluating their influence on the final geometry of the sample.

As comparison methods, height and width measurements were taken for each deposited single bead and wall, according to Fig. 1, representing the measurement points for the walls, spaced at 10 mm intervals between each point. For the single beads, the distance between the points was 5 mm, due to higher variation along the length of these samples. For each measurement point, 5 measurements were taken and the mean with standard deviation was calculated (with a confidence interval of 99.70%). After the external measurements were verified, the cord/wall was cut in half and the total and effective width in the section were measured.

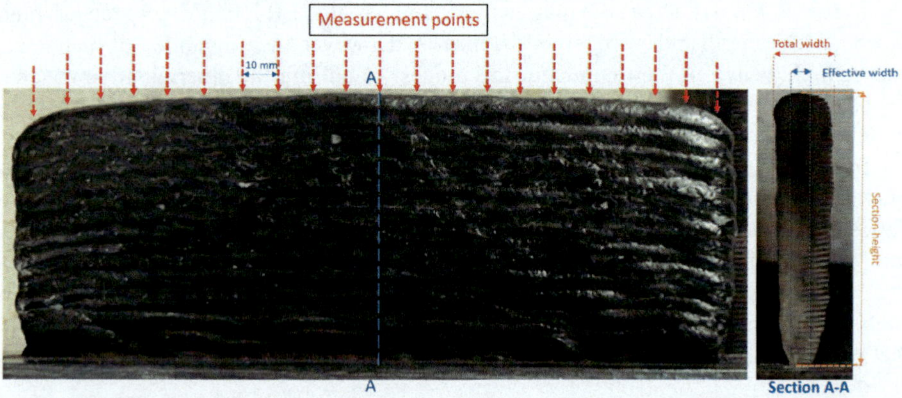

Fig. 1. Measurement points schematic for the samples deposited through the length and cross-section ate the middle of the sample (Section A-A).

Through measurements, the dimensions of the samples were compared to determine the configuration used for a deposition that resulted in the highest height and narrowest width of the bead. Additionally, the effective areas in the sample sections were evaluated in order to determine those that achieved the highest material utilization. The walls of the samples were compared to those obtained through conventional GMAW deposition [8] to verify the improvements achieved by the CMT process in terms of geometry. Furthermore, a visual inspection was performed in accordance with ISO standard 5817 [9] to analyze any visually determined discontinuities/defects.

The energy values were also compared between the CMT and conventional GMAW processes, as stated by Fronius International GmbH the CMT process requires less energy to generate and stabilize the arc when compared to conventional GMAW processes. To this end, power values of the sources and the effective thermal input applied to the workpiece, corrected by a thermal efficiency for each process, were calculated. For the comparison, the walls with 15 and 30 layers were evaluated, and the walls with 16 layers manufactured by Santos [8] via conventional GMAW were used for comparison.

3 Results

3.1 Preliminary Depositions

For geometry analysis, 7 simple beads were deposited, where three of which are presented in Fig. 2. The objective of the preliminary depositions was to verify the influence of torch displacement speed and wire feed speed on the geometry of the samples. For the displacement speed of 4 mm/s combined with wire feed speed from 1.7 m/min to 3 m/min, the samples presented a pre-humping or humping effect. Yuan et al. [10] correlate this effect as a consequence of high displacement speed or low heat input required for the molten pool stabilization, necessary for obtaining regular geometries for the samples, without several discontinuities.

The deposition most affected by the humping effect was the bead represented in Fig. 2a, which geometry indicates the lack of flow by the molten pool in the beginning, increasing by the final of length due to the heating of the substrate material. The sample presented in Fig. 2b also presented the initial effect (pre-humping), as indicated by the red arrows.

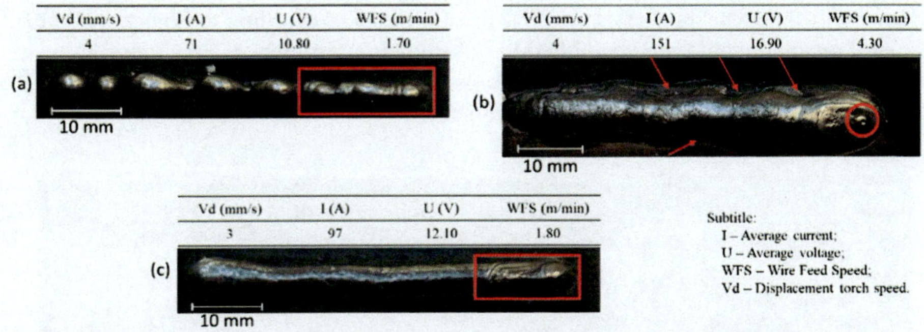

Fig. 2. Preliminary simple beads deposited: (a) sample presenting humping effect with simple improvement at final length (red rectangle); (b) sample with pre-humping effect, indicated by red arrows, and crater pipe, indicated at red circle; (c) sample with best geometry and the presence of material retraction (red rectangle).

A similar scenario is related by Chaudhari et al. [11] for the deposition of 316L stainless steel, as the authors verify that the high torch displacement speed is the main responsible for the effect. Therefore, in order to avoid this effect on the sample walls depositions, it was decided to apply higher wire feed speeds, alongside a lower torch displacement speed, aiming to elevate the heat input of the process.

Another feature presented in 3 of 7 bead samples for the preliminary stage was the crater pipe, defined by ISO 5817 [9], highlighted in Fig. 2b. This effect is generated by the lack of material fill with the retraction presented by the bead solidification. However, the crater pipe presented in the samples did not present superficial cracks, and according to the standard, the effect presented is not considered a formal discontinuity.

The sample which presented the best geometry for the preliminary group is presented in Fig. 2c. The sample was deposited using a torch displacement speed of 3 mm/s and wire feed speed of 1.8 m/min, allowing to obtain a sample free of defects and discontinuities, with a width of 3.57 ± 0.05 mm and height of 5.07 ± 0.05 mm. The final length presented a material retraction during cooling (highlighted in Fig. 2c). However, the effect was not considered a former discontinuity and it is not relevant for multiple layers depositions because of the reheating for each layer deposited. This configuration was chosen as the initial configuration for the 5-layer wall depositions.

3.2 Wall Depositions

This stage consisted of the deposition of 5, 10, and 30-layer walls, once which group parameters were based on the results from the previous group deposited. Firstly, 4 samples of 5-layer walls were deposited based on the parameters used for the sample in Fig. 2c. The stage also introduced the step-over layer distance, which indicates the torch height displacement from the substrate. The contact tip-to-work distance (CTWD) was used as a reference, and the initial distance was defined as 12 mm for the CMT-WAAM process, according to Borovkov et al. [12] to maintain arc stability during deposition. The 5-layer samples presented a significant height difference between the center and the borders of the wall, with a height difference near to 30% higher in the middle of the wall when compared to the borders, as represented in Fig. 3. According to Zhang et al. [13], one of the methods that can be applied to reduce height difference is the interruption of arc between layers in order to control the viscosity of the deposited material along the length and avoiding high heat concentration at the borders due to torch trajectory.

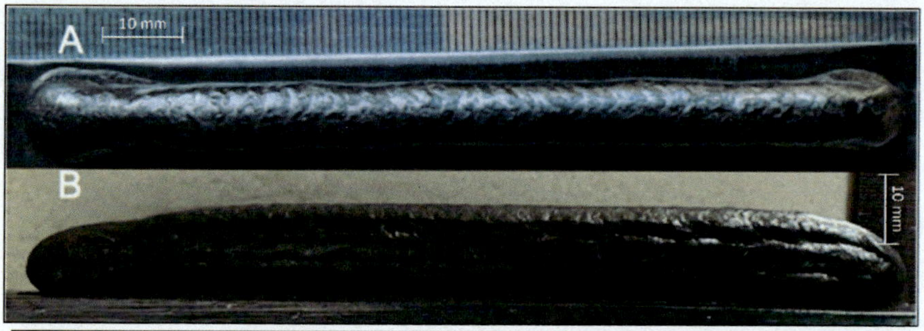

Vd (mm/s)	Step-over distance (mm)	Final CTWD (mm)	I (A)	U (V)	WFS (m/min)
3	1	4.50	112	11.60	3

Fig. 3. 5-layer wall with deposition parameters, presenting height difference between the center and borders, showed in superior view (A) and lateral view (B).

In addition to the proposition of Zhang et al. [13], an alternative to avoid the height reduction at the borders is the use of a higher displacement speed to reduce heat input at the regions. Using a speed range from 4 mm/s to 6 mm/s, the width of the borders, which used to be larger due to the heat concentration, was reduced by 13% with the change.

Another aspect that influences the heat concentration is the reduction of CTWD along the process, reducing arc length and, consequently, generating a lower arc contact area. This effect was registered for the first two samples deposited with 1 mm of step-over distance, presenting at the end of the depositions 4.5 mm and 6 mm of CTWD. Therefore, the other two samples from the 5-layers wall group applied a 2 mm step-over distance, using displacement speeds of 4 mm/s and 6 mm/s. Both samples resulted in a small variation of CTWD, with 10 mm and 11 mm respectively at the end of the depositions.

The sample that resulted in the best geometric characteristics for this group is presented in Fig. 4, with the mean width reduced compared to others, which indicates higher molten pool control, and a mean height of 8.89 ± 0.07 mm. Despite this sample presenting the lower height variation along all the length, the difference was stated in 22%.

For the 15-layer walls group, the first sample were submitted by a step-over distance of 2 mm, and the parameters applied for the sample shown in Fig. 4. However, the configuration resulted in a higher CTWD of 20 mm, which can be harmful to arc stability, especially for depositions with a greater number of layers. Through these observations, the second sample used a step-over distance of 1.5 mm, leading to a final CTWD of 13 mm, closer to the initial value of 12 mm. The sample also resulted in a lower height variation compared with the previous samples, as presented in Fig. 5.

Vd (mm/s)	Step-over distance (mm)	Final CTWD (mm)	I (A)	U (V)	WFS (m/min)
4	2	10	89	11.80	2.60

Fig. 4. 5-layer wall deposited that presented the best geometric characteristics, alongside the deposition parameters. Superior view (A) and lateral view (B).

The height variation is also noticeable for the samples in Fig. 5, as the effect is more expressed for sample 1 due to the higher CTWD and, consequently, a higher arc instability, verified by the higher voltage resulting from the deposition, as compared with sample 2. This height difference is mainly caused by the material flow due to the higher molten pool, generating a buildup of material on the sides of the wall. The second sample, which obtained an adjusted CTWD, presented a mean width of 7.46 ± 0.07 mm, while the first sample presented an 8.29 ± 0.13 mm width. The height variation stayed below 5% for the second sample, lower than the first sample that presented approximately 10%.

The side slope presented in both samples in Fig. 5 can be solved using interpass temperature control or deposition control methods, such as electromagnetic oscillation used by Corradi et al. [14]. The authors applied coils and magnetic amplifiers for the flow material control by Lorentz force that acts in the molten pool. The method resulted in better geometry quality for the samples deposited and proved to be an alternative for quality control.

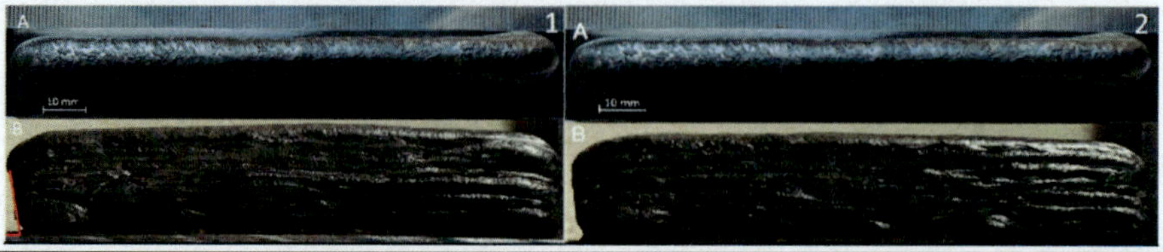

Sample	Vd (mm/s)	Step-over distance (mm)	Final CTWD (mm)	I (A)	U (V)	WFS (m/min)
1	6	2	20	67	18.20	2.10
2	6	1.50	13	87	12.20	2.70

Fig. 5. 15-layer walls deposited with the parameters used for deposition.

For the 30-layer walls group, the samples were separated in three different deposition methods: continuous deposition, with a non-interrupt arc as the previous samples were conducted, deposition by steps, with steps with 15 layers deposited, followed by temperature control of 100ºC using natural cooling at 25 ºC room temperature and deposition with temperature control for each layer deposited, using natural convection cooling until a temperature range of 45ºC to 55ºC. For each strategy, the geometric characteristics were checked to determine the individual aspects of each one. The temperature ranges were determined based on deposition cycles from larger parts, as an example the MX3D bridge constructed with GMAW process [15].

The sample deposited by steps, divided into 15 layers each, did not present any discontinuities or visible defects, except for the middle section where occurred the stop for cooling. The area presented a visible reduction in width (see Fig. 6), resulting in a lower effective area. According to Lehmann et al. [16], the use of an interlayer temperature control method reduces the variation of each layer's thickness, once the cooling for each layer generates a homogeneous heat input for each layer, maintaining the molten pool region. However, natural cooling also increases manufacturing time, turning this cooling method into a barrier to the productivity required for industrial applications.

The sample fabricated via continuous deposition presented a regular geometry, without any significant disparities and with effective width near to the total width of the sample. When compared with the sample deposition by steps, the mean height presented close values, with 36.84 ± 0.53 mm for the continuous deposition, and 37.16 ± 0.66 mm for the deposition by steps. Also, the width values presented similar values, with 8.42 ± 0.10 mm and 8.76 ± 0.10 mm respectively.

The deposition using interlayer temperature control presented a mean height 25% higher than other samples for the group, even using the same initial deposition parameters, with 45.88 ± 0.39 mm. The mean width resulted for the sample was also 30% lower than other walls, with 6.16 ± 0.08 mm, as shown in Fig. 7. The interlayer cooling method was also applied by Lehmann et al. [16], which authors related similar improvements in the geometry of the layers for mild steel alloys.

3.3 Cross-Section Analysis

After the deposition of the samples, all of them were sliced at the middle of samples length in order to visualize the cross-section geometry of the samples and calculate the

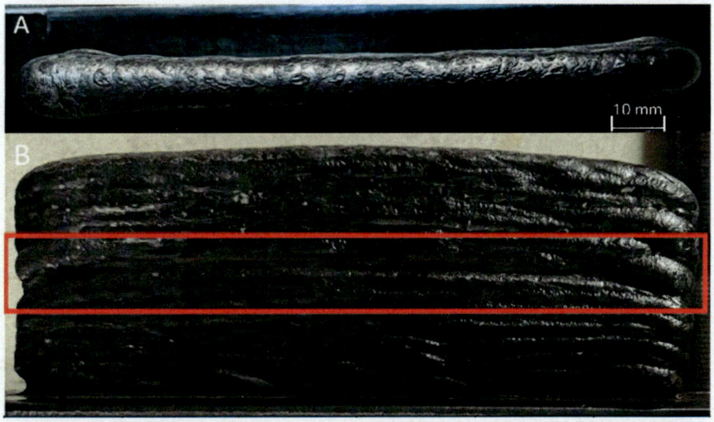

Vd (mm/s)	Step-over distance (mm)	Final CTWD (mm)	I (A)	U (V)	WFS (m/min)
6	1.50	15	112	11.60	3

Fig. 6. 30-layer wall deposited by steps (15 + 15 layers) with deposition parameters. The layer thickness difference for the middle section is highlighted.

Fig. 7. 30-layer wall deposited using interlayer temperature control.

total and effective measurements. The width and height of the walls and beads. The reinforcement and contact angle were also measured for the beads, with the sections of samples from all groups presented in Fig. 8.

For the simple beads group, the sample from Fig. 2c presented a contact angle of 81°, the closest to achieving 90° wished for the wall depositions. The sample presented a thickness of 4.68 mm and reinforcement of 3.67 mm. For the 5-layer walls group, it was verified that a higher displacement speed generated a reduction of the effective area of the section, with the lowest ratio between the total and effective area of the group. The sample from Fig. 4 presented the highest ratio from the group, with an effective area of 87% of the total area, while the other sample presented an 83% ratio.

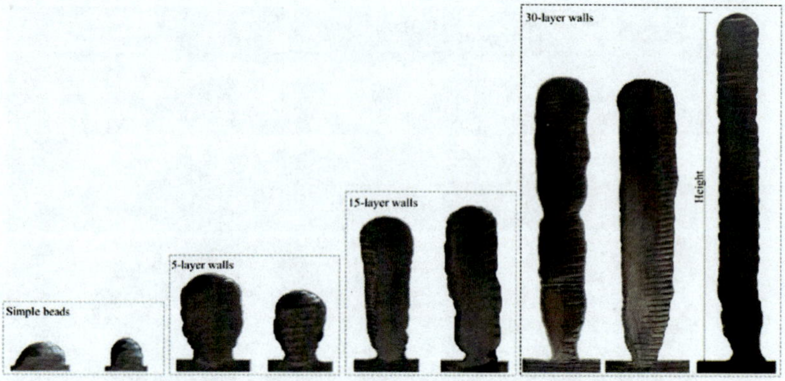

Height (mm)	3.14	3.67	11.30	9.12	19.15	20.44	38.77	38.16	45.19
Cross-section	Fig. 2b	Fig. 2c	Fig. 3	Fig. 4	Fig. 5 Sample 1	Fig. 5 Sample 2	Fig. 6 Step deposition	Continuous deposition	Fig. 7 Interlayer control deposition

Fig. 8. Parallel between all samples presented in Figs. 2, 3, 4, 5, 6 and 7.

For the 15-layer group, it is noticeable that the second wall from Fig. 5 did not present a good linearity along the height, which can contribute to material collapse for higher depositions. Despite it, the sample presented a higher effective area ratio, with 87% compared to 83% from the first wall. The linearity aspect was also mentioned by Arana et al. [17], who attributed the effect to a high thermal input, causing a lower viscosity of the material leading to slipping and possible collapse of the wall or part of it.

The 30-layer walls with both continuous and step depositions resulted in similar heights and widths, with less than 5% difference, even considering the reduction in mid-height from the step method wall. When compared with the wall deposited with interlayer temperature control, the height variation was near 20%, as the width presented 5.2 mm for the interlayer method, as the others had more than 7 mm. The area ratio obtained by the interlayer method was 86%, 6% higher than step method wall, and 14% higher than the continuously deposited wall.

Besides the benefits that interlayer temperature control brings to sample geometry, the natural cooling process turns its practice unfeasible. The alternative is the application of active cooling methods that reduces manufacturing time. Da Silva et al. [18] developed an active cooling method using a variable water vat named Near Immersion Active Cooling. The process consists in submerge the deposited material in order to dissipate all heat and keep the parts homogeneous along all the manufactured part. This method allowed depositing a 29-layer wall in 11 min, compared with the same geometry deposited in approximately 4 h with natural cooling, representing a 95% time saving to achieve similar results.

3.4 Heat Input

This section has the objective to compare CMT process used in this work with conventional GMAW deposition process applied by Santos [8] using the ER410NiMo material. For the CMT process, it was determined that the average heat input registered by RCU 5000i Fronius controller had a thermal efficiency of 0.85, recommended by Lee [19]. The 15-layer walls, deposited using a displacement speed of 6 mm/s, resulted in a heat input of 0.161 kJ/mm. Santos [8], using a displacement speed of 12 mm/s and thermal efficiency for GMAW of 0.78, determined a heat input of 0.165 kJ/mm for a 16-layer wall sample. Considering the same power output applied by conventional GMAW applied for a displacement speed of 6 mm/s for comparison purposes, the heat input would increase to 0.330 kJ/mm, approximately 104.96% higher than CMT. Table 1 presented the average values of heat input and power output for the 15 and 30-layer samples compared with Santos [8] samples deposited by conventional GMAW.

Table 1. Average power output and heat input values resulted from the CMT and conventional GMAW samples.

Samples	15-layer walls	30-layer walls	Average CMT samples	Average from the GMAW samples for Vd = 12 mm/s [8]	Average from the GMAW samples for Vd = 6 mm/s [8]
Thermal efficiency	0.85			0.78	
Electric power output (kW)	1.14	1.25	1.19	2.53	
Heat input (kJ/mm)	0.161	0.180	0.171	0.165	0.330

4 Conclusions

The geometry study of the simple beads and walls fabricated with stainless steel ER410NiMo by different deposition parameters and methods allowed to obtain the following conclusions:

- The humping effect was presented at simple bead deposition when applying wire feed speed from 1.7 m/min to 3 m/min, using a displacement torch speed of 5 mm/s.
- The best configuration applied for the simple beads, was with a 3 mm/s displacement speed and 1.8 m/min of wire feed speed, resulted in a contact angle of 90° and a reinforcement approximately 16% higher than other samples from the group.

- The use of a 6 mm/s torch displacement speed and a wire feed speed of 3 m/min for the 5-layers wall group resulted in a 22% width reduction for a 2 mm step-over distance, acting as the main factor for the more stable height along the length as well.
- The step-over distance was a key factor for good geometry characteristics, as the heat-affected zone is highly influenced by the arc contact area ruled by the parameter, as observed for the 15-layer walls group.
- The wall deposited by step deposition method with a cooling temperature of 100ºC resulted in a height 1.6% higher than the continuously deposited sample. However, the sample also presented a reduced effective width due to the cooling.
- The sample deposited using interlayer temperature control method resulted in a 86% effective ratio, compared the total and effective area. Furthermore, the cooling closer to room temperature resulted in a more linear sample 25% higher in height and with a 6.16 ± 0.08 mm average width.
- The heat input applied for the CMT samples were 51.21% lower than GMAW compared with the same displacement speed, the CMT also presented more regular geometry.

Acknowledgments. The authors would like to thank the Department of Mechanical Engineering (ENM), Faculty of Technology (FT), Post Graduate Program in Mechanical Sciences (PCMEC) at the University of Brasilia (UnB), the Research Support Foundation of the Federal District (FAPDF) and CAPES for support and laboratory/financial support. The authors are also thankful to CNPq (process 405499/2022–1).

References

1. Precedence research, Homepage. https://www.precedenceresearch.com/metal-additive-manufacturing-market. Accessed 19 Feb 2023
2. Paolini, A., Kollmansberger, S., Rank, E.: Additive manufacturing in construction: a review on processes, applications, and digital planning methods. Addit. Manuf. **30**, 100894 (2019)
3. Wu, B., et al.: A review of the wire arc additive manufacturing of metals: properties, defects and quality improvement. J. Manuf. Process. **35**, 127–139 (2018)
4. Selvi, S., Vishvaksenan, A., Rajasekar, E.: Cold metal transfer (CMT) technology - an overview. Def. Technol. **14**, 28–44 (2018)
5. Cornacchia, G., Cecchel, S., Panvini, A.: A comparative study of mechanical properties of metal inert gas (MIG)-cold metal transfer (CMT) and fiber laser-MIG hybrid welds for 6005A T6 extruded sheet. Int. J. Adv. Manuf. Technol. **94**, 2017–2030 (2017)
6. Novelino, A.L.B., Carvalho, G.C., Ziberov, M.: Influence of WAAM-CMT deposition parameters on wall geometry. Adv. Ind. Manuf. Eng. **5**, 100105 (2022)
7. Bruno, J.M.A.: WAAM processing of AISI 410 stainless steel for part building, M.Sc. Thesis, University of Lisbon, Higher Technical Institute, Portugal, pp. 1–79 (2019)
8. Santos, T.M.: Estudo do Efeito da Temperatura de Interpasse e de Pré-aquecimento no Processo de Soldagem por Múltiplos Passos no Reparo de Pás de Turbina Hidráulicas, pp. 1–164, Ph.D. Thesis, University of Brasilia, Brazil (2018)
9. ISO 5817.: Welding - Fusion-welded joints in steel, nickel, titanium and their alloys (beam welding excluded) - Quality levels for imperfections. International Standard, pp. 1–26 (2023)
10. Yuan, L., et al.: Investigation of humping phenomenon for the multi-directional robotic wire and arc additive manufacturing. Robot. Comput.-Integr. Manuf. **63**, 101916 (2020)

11. Chaudhari, R., Parmar, H., Vora, J., Patel, V.: Parametric study and investigations of bead geometries of GMAW-based wire–arc additive manufacturing of 316L stainless steels. Metals **12**(7), 1232 (2022)
12. Borovkov, H., et al.: In-line height measurement technique for directed energy deposition processes. J. Manuf. Mater. Process. **5**(3), 85 (2021)
13. Zhang, Y., Chen, Y., Li, P., Male, A.: Weld deposition-based rapid prototyping: a preliminary study. J. Mater. Process. Technol. **135**(2–3), 347–357 (2003)
14. Corradi, D., Bracarense, A., Wu, B., Cuiuri, D., Pan, Z., Li, H.: Effect of magnetic arc oscillation on the geometry of single-pass multi-layer walls and the process stability in wire and arc additive manufacturing. J. Mater. Process. Technol. **283**, 116723 (2020)
15. MX3D Homepage, MX3D bridge. https://mx3d.com/industries/mx3d-bridge/. Accessed 13 Feb 2023
16. Lehmann, T., et al.: Concurrent geometry- and material-based process identification and optimization for robotic CMT-based wire arc additive manufacturing. Mater. Des. **194**, 108841 (2020)
17. Arana, M., Ukar, E., Rodriguez, I., Aguliar, D., Alvares, P.: Influence of deposition strategy and heat treatment on mechanical properties and microstructure of 2319 aluminium WAAM components. Mater. Des. **221**, 110974 (2022)
18. Da Silva, L.J., Souza, D.M., de Araújo, D.B., Reis, R.P., Scotti, A.: Concept and validation of an active cooling technique to mitigate heat accumulation in WAAM. Int. J. Adv. Manuf. Technol. **107**, 2513–2523 (2020)
19. Lee, S.: CMT-based wire arc additive manufacturing using 316L stainless steel: effect of heat accumulation on the multi-layer deposits. Metals **10**(2), 278 (2020)

Impact of Deposition Procedures on 410L Multilayer Plasma Transferred Arc Processing

Otavio de Oliveira Lima[1,2(✉)], Gustavo Scheid Prass[1], and Ana Sofia C. M. d'Oliveira[1]

[1] Department of Mechanical Engineering, Federal University of Paraná, Av. Cel. Francisco H. dos Santos, 100 – Jardim das Américas, Curitiba, PR, Brazil
{otavio.lima,gustavo.prass,sofmat}@ufpr.br
[2] Volvo Construction Equipment North America, LLC. 304 Volvo Way, Shippensburg, PA, USA

Abstract. Additive manufacturing (AM), Directed Energy Deposition (DED) technology is an important industrial manufacturing tool to produce mechanical parts without using expensive molds with functionalities otherwise not possible. AM can be carried out to fabricate new components or on maintenance to rebuild the geometry of worn parts. Plasma Transferred Arc (PTA) is a low-cost, high-quality deposition technique that brings significant advantages in AM, and is used to process multilayers of AISI 410L. This study is part of an on-going study that addresses the impact of the deposition direction, bidirectional and unidirectional, and the mass flow (Qm) on the microstructure's solidification of multilayers. Walls were built using four processing conditions and characterized in regions exposed to a larger and a small number of thermal cycles. Results pointed out that regions exposed to many thermal cycles, exhibited larger ferritic grains with an acicular structure at the grains boundaries and a uniform hardness profile. In contrast, the top region of multilayers exposed to a smaller number of thermal cycles revealed changes to the microstructure and hardness profile. The later exhibiting lower values at the last deposited layers up to 5 thermal cycles where larger grains formed followed by an increase after 8 thermal cycles and the onset of the acicular structure. The low mass flow used allow for good finishing of the multilayers nevertheless the increase in mass flow resulted on an increase in hardness at the different regions and more significant changes at the top layers.

Keywords: Additive Manufacturing · Martensitic Stainless Steel · Plasma Transferred Arc · Quenching and Tempering

1 Introduction

Technological advances available for the productive sector and its processes, added to the need for them to be increasingly sustainable, are considered a differential for companies operating in the manufacturing sector. Among the existing processes, the most used are casting and machining. The first presents a high cost due to the need to manufacture molds that are used only once or by high resistance molds for continuous use. In machining, either the conventional or CNC, the material removal process demands high-energy

D. de Oliveira et al. (Eds.): COBEF 2023, LNME, pp. 126–139, 2024.
https://doi.org/10.1007/978-3-031-43555-3_12

consumption and generates a high amount of waste and, on some occasions, does not allow the production of parts with complex geometries.

Additive Manufacturing emerges as an innovative process that enables to manufacture parts and components with customized geometry, being able to offer complex geometries and with reduced costs when compared to conventional manufacturing processes through printing part by part, layer by layer until obtaining the final geometry defined in a 3D software (Thompson et al. 2016).

In this context, one of the benefits of AM stands out: the ability to use materials effectively, using the necessary amount for printing the component, reducing waste, and showing great relevance especially when working with materials that have a high-cost acquisition, considering the lowest need to subtract material (Pereira et al. 2018).

AM technology, particularly DED processes, allows to both designing and manufacturing parts directly from a digital file without the need of expensive tooling and the ability to add functionalities including structures with multiple materials. Within DED processes existing challenges and needs for the development of AM parts include planning the trajectory for the deposition of the desired part is highly relevant, however, the lack of solutions or alternatives for its optimization has been the subject of discussion for most of the problems related to the AM (Liu and To 2017).

This investigation addresses this challenge assessing the impact of the deposition direction (bidirectional and unidirectional) and the powder-flow rate (6 g/min and 9 g/min) used on the microstructure and hardness of multilayers of atomized AISI 410L stainless steel.

2 Materials and Methods

Table 1. Deposition parameters.

Parameter	Value
Protective gas	Argon 99%
Plasma gas flow (l/min)	2.00
Shield gas flow (l/min)	15.00
Carrier gas flow (l/min)	0.80
Nozzle-plate distance (mm)	10
Deposition current (A)	150
Deposition speed (mm/min)	150
Powder flow rate (g/min)	6 and 9
Deposition direction	Bidirectional and Unidirectional
Deposition starting temperature between the multilayers (°C)	150

Multiple layers of gas atomized AISI 410L stainless steel powder (53–150 μm) were deposited using Plasma Transferred Arc (PTA) on AISI 304L substrate plates (L = 150.0 mm, W = 12.7 mm, H = 31.7 mm) to obtain a 20 mm height multilayer wall. Four sets of processing parameters were used, Table 1, to assess the impact on microstructure and hardness depending on the processing conditions. Figure 1 shows the flowchart of multilayer deposition process and their characterization.

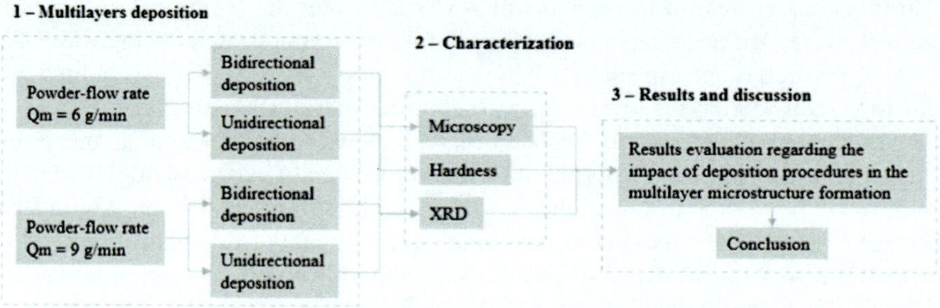

Fig. 1. Flowchart of multilayer deposition process and their characterization.

Figure 2 illustrates the deposition strategy according to the selected direction. An interlayer temperature was set at 150 °C, measured at the center of each deposited layer. During the bidirectional processing, from position A to B when the center region reaches 150 °C the deposition of the second layer starts from position B to A, the procedures is repeated during the build-up of the multilayer wall. Similar procedure was used for the unidirectional processing except that in all layer's deposition was carried out from position A to position B. The starting deposition temperature for bidirectional multilayers is higher than that of unidirectional multilayers because the deposition starting point was allowed more time to cool. Air-cooling to the interlayer temperature depended on the deposition strategy, with an average air-cooling time of 367 s for the bidirectional walls were while for the unidirectional walls were 515 s.

For the four set of processing conditions, analysis was carried out at two regions associated with different thermal cycles: the center region, which has a higher thermal cycle exposition, and the top region, exposed to a lower number of thermal cycles. Walls built with 6 g/min powder flow rate had a total 35 thermal cycles while the 9 g/min walls had 25 thermal cycles at maximum, with an average thickness of 0.65 mm and 0.98mm, respectively.

Five mm thickness samples were removed at the center of multilayer walls processed with the four set of parameters and embedded, followed by grinding with silicon carbide, and polished with 1 μm alumina. Polished samples were etched with Vilella's reagent (10 ml of hydrochloric acid, 4 g of picric acid and 200 ml of alcohol) for 1 min. The Vilella's reagent reveals the ferritic structure (α) etching the grain boundary (Silva et al. 2005), the second phase particles (carbides, σ phase and δ ferrite) and reveals the martensite (Voort et al. 2004).

The hardness measurements were performed at the cross section of the samples in the center and top regions. For each region, measurements were made in a 5 x 3 matrix

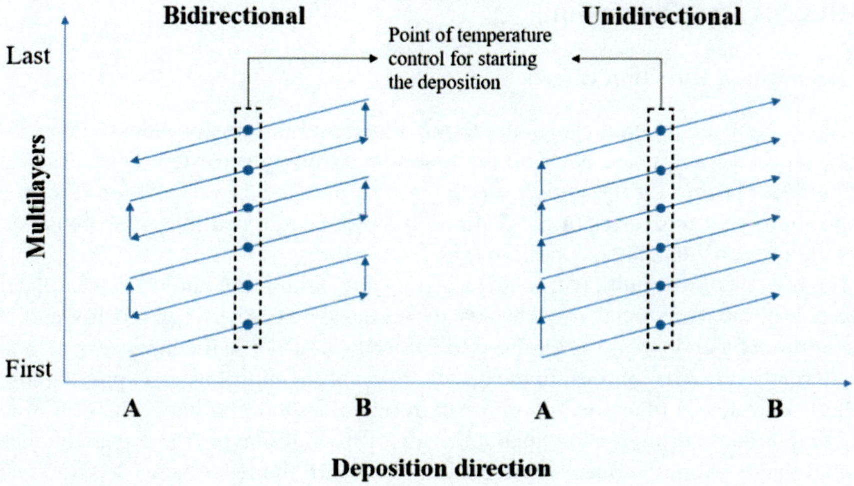

Fig. 2. Deposition strategy.

configuration, Fig. 3. On the top region, the upper measurements were taken 1.5 mm away from the surface and all indentations are 1 mm apart horizontally and vertically. The hardness measurements followed the ASTM E384/2017, Standard Test Method for Microindentation Hardness of Materials and the used parameters were 10 N load and indentation time of 30 s.

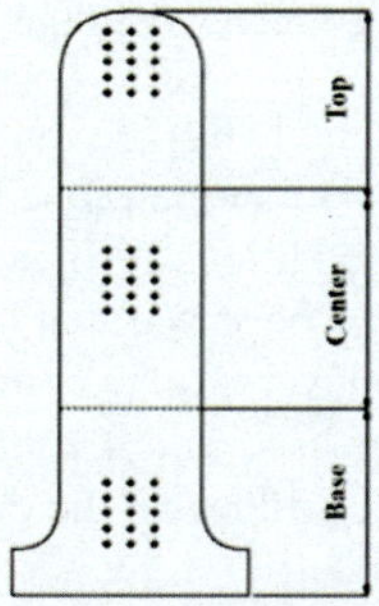

Fig. 3. Hardness measurement matrix.

The as-deposited multilayers underwent to X-ray diffraction analysis with 2θ interval between 40 and 120 degrees, scanning speed of 2°/min and Cu tube. The results were analyzed with Chrystallografica Search-Match software.

3 Results and Discussions

3.1 Deposition Direction Effects

The analysis of the thermal cycles in the processed multilayers for the two deposition directions used was carried out with the lowest mass flow rate tested, Qm = 6 g/min. The hardness profile for the multilayers processed with the two deposition directions, Fig. 4, shows that regardless of the deposition direction adopted, the same behavior is observed to the as-deposited condition.

Top regions of the multilayers show a significant variation in hardness. Near the top surface, exposed to up to 5 thermal cycles, the hardness varies between 155 HV and 185 HV, followed by an increase in hardness to 210 HV - 230 HV in the layers exposed to 8 or 9 thermal cycles. In contrast, in the center region of the multilayers, exposed to up to 25 thermal cycles, hardness exhibit more uniform values varying between 180 HV and 200 HV. Hardness profiles show that at the top of the multilayers there is a volume of material that is strongly affected by the thermal cycle of the layer being deposited, and in regions far from the top, the volume of material exhibits uniform hardness, suggesting that the deposition temperature no longer induces significant transformations in the AISI 410 stainless steel.

The last deposited layers exhibited a similar hardness profile a consequence of the solidification microstructure response to up to 7 heating and cooling cycles. However, walls processed with a unidirectional deposition strategy shifts the peak hardness to a smaller number of cycles also near the top of the wall hardness is lower when compared to measurements made on near the top of wall build with a bidirectional scanning strategy. The observed differences in the hardness profile are associated with the starting deposition temperature of each layer and subsequent slower cooling rate measured for the unidirectional strategy.

The measured differences associated with the direction of deposition are reinforced by the work of Nikam and Jain (2017) that mentioned that the temperature gradient for each multilayer deposited and the conditions for heat diffusion are better in the unidirectional deposition. The authors also observed that temperature gradient was constant after the deposition of the top five layers, reinforcing the measured hardness profile, Fig. 4. For the processing conditions used in our study peak hardness measured on the top region of multilayers is similar regardless of the deposition strategy as it is hardness measured at the center of multilayers.

The different deposition directions used to build the multilayer walls of AISI 410L impact in the formation growth of the microstructure as represented in Fig. 5. The bidirectional deposition of multilayers induces a zigzag effect in the microstructure growth while unidirectional multilayer deposition is expected to impose the same growth orientation for each deposited layer of the processed wall. This behavior is not expected to be as pronounced as that observed in Powder Bed Melting (PBF) processes, which form a long and shallow melt pool that strongly affects grain orientation and structure (Dogu et al. 2022).

Epitaxial growth, which takes advantage of the crystalline orientation of the grains from the previous layer, is observed from the first deposited layer in the multilayers regardless of the deposition directions. This phenomenon has also been described by

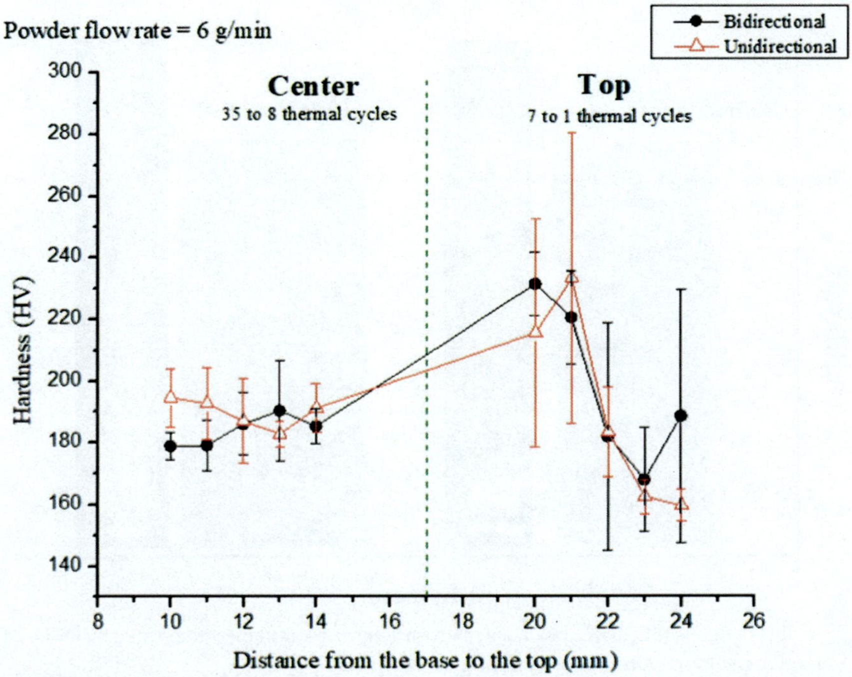

Fig. 4. As-deposited multiplayers hardness profile with 6 g/min of powder flow rate.

Yehorov, da Silva and Scotti (2019) who also addressed microstructure growth with different deposition directions using wire of IN718. Parimi et al (2014) describe that in the unidirectional deposition columnar grains are smaller, and the growth direction exhibits an angle of less between 50° and 60°, while the bidirectional deposition results in angles between 90° and 100° relative to the substrate, and its inclination follows the back of the weld pool. The authors mention that this orientation could be related to the influence of vertical and horizontal heat flow, respectively, as well as the dendritic orientation of previously deposited layers.

The thermal history that affected the measured hardness is also responsible for the microstructure of the multilayers. Figure 6 shows the longitudinal and transversal cross section of the microstructure of the multilayers processed with both bidirectional and unidirectional deposition strategy using 6 g/min as powder flow rate. It is observed that when the bidirectional deposition is adopted, it results in long columnar grains from the substrate to the top layers that in the cross section exhibit a radial growth orientation in relation to the substrate. In contrast, in the unidirectional deposition, grains at the transverse cross section are shorter corresponding to a section of the long columnar and incline grains observed in the longitudinal direction of the processed walls.

Regardless of the deposition direction, it is observed that epitaxial growth occurs during solidification in each deposited layer of the walls. This behavior is expected when layers with the same chemical composition and crystalline structure are deposited by welding processes, as cited by (Alberti et al. 2015). In the top region, grain size is

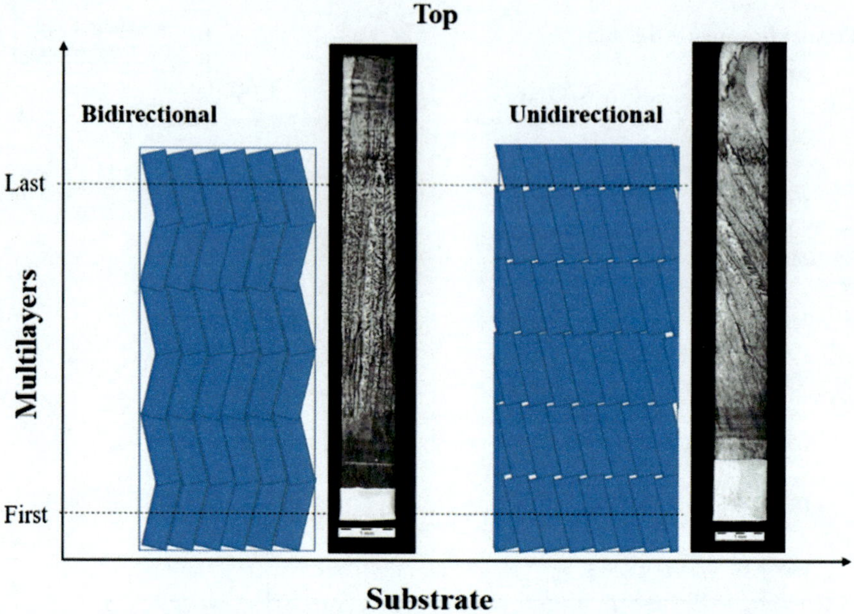

Fig. 5. Effect of deposition direction on the microstructure of multilayers deposited by DED-PTA, as observed in the longitudinal section of processed walls.

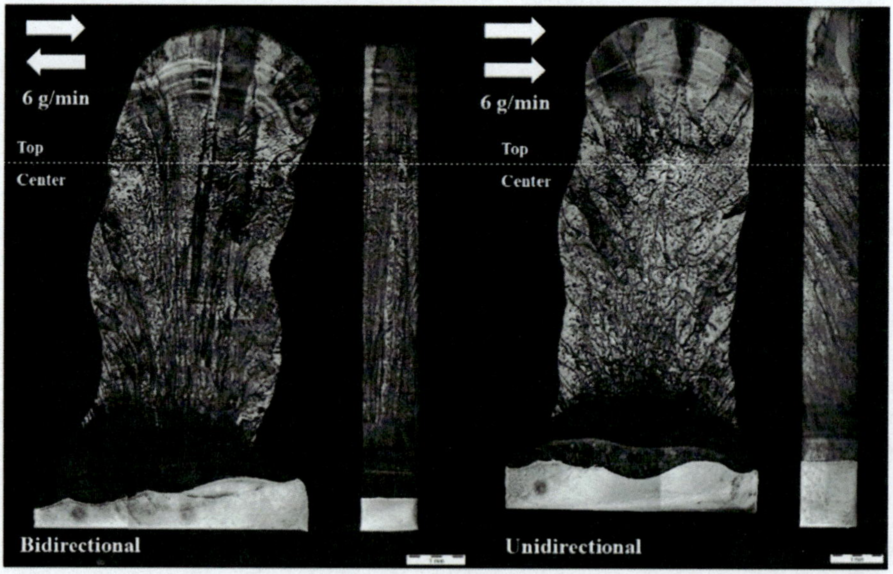

Fig. 6. Bidirectional and Unidirectional walls microstructure deposited with 6 g/min powder flow rate.

larger when compared to the center and base regions due to the high temperature at the top layers and smaller number of thermal cycles that were not enough to induce significant transformations in the solid state (Khodabakhshi et al. 2020).

Figure 7 shows details of the microstructure for the as-deposited multilayers processed with 6 g/min in the bidirectional and unidirectional directions. Both processing conditions result in large grains covering the last 2–3 deposited layers exposed to more severe cooling-rate. For 410L steel, which has a high Cr/Ni ratio, it is expected that the first phase to solidify to be the δ ferrite and the high cooling rates associated with both PTA process and multilayer deposition induce a greater ferrite formation, since the transformation in the solid state of ferrite into austenite takes less time to occur (Kou 2003).

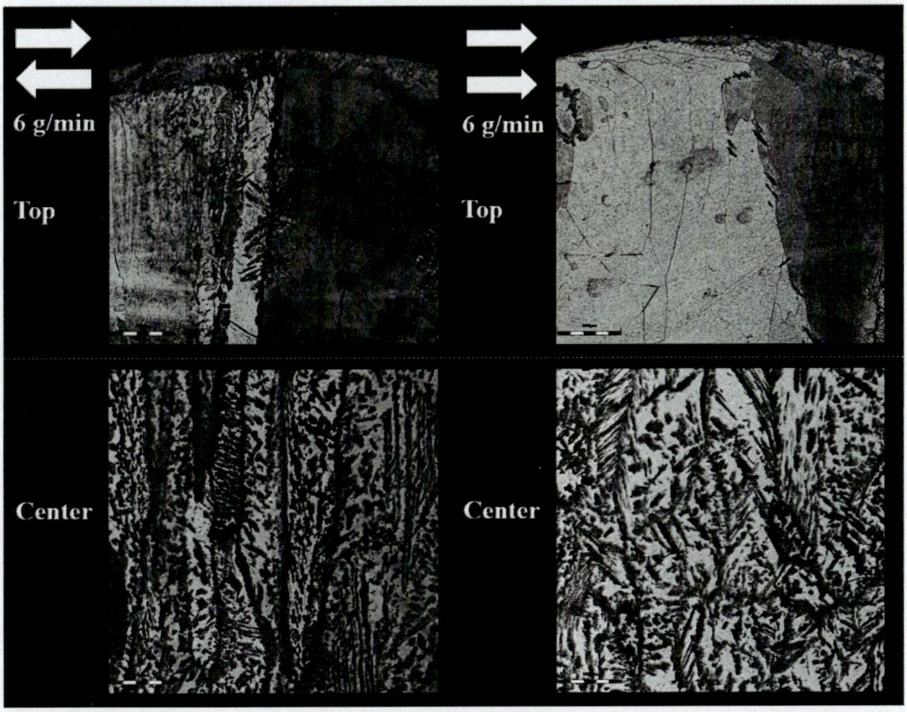

Fig. 7. Cross-section microstructure detail for bidirectional and unidirectional depositions with 6 g/min powder flow rate with 500x enlargement.

At the top of the multilayers an acicular structure forms at the grain boundaries identified as Widmanstatten austenite associated with the transformation in the solid state of ferrite → austenite exposed to high cooling speeds. Kou (2003) also highlights the possibility of formation of either lacy or vermicular ferrite during the rapid solidification of steels with ferritic matrix, which is compatible with the observations made in the cross-section microstructure of the two deposition directions used.

For a better understanding of the transformations described above, it is necessary to point out that the martensitic and ferritic phases predicted by the Schaffler and DeLong diagrams for AISI 410 steel, which agrees with the diffractogram made in the cross section, considering that the ferrite and martensite peaks are the same. This result can be associated with the cooling rates to which the multilayers were exposed, and the low carbon content of the steel used, AISI 410L.

Also, according to the Fe-Cr phase diagram and for the Cr content present in the alloy, the formation of austenite occurs after the formation of primary ferrite (δ). It is also relevant to highlight that, for alloys with low carbon content, the austenite loop is reduced, expanding the two-phase region, which tends to favor austenite, which will have Widmanstatten characteristics when cooled rapidly. The incoherent curved interfaces of the acicular phase have high mobility and allow the fast growth and continuous advancing on the ferrite. In the central region, exposed to the highest number of thermal cycles, a structure of ferritic grains and evidence of martensite is identified, as shown by the acicular structures that cut the grains transversally.

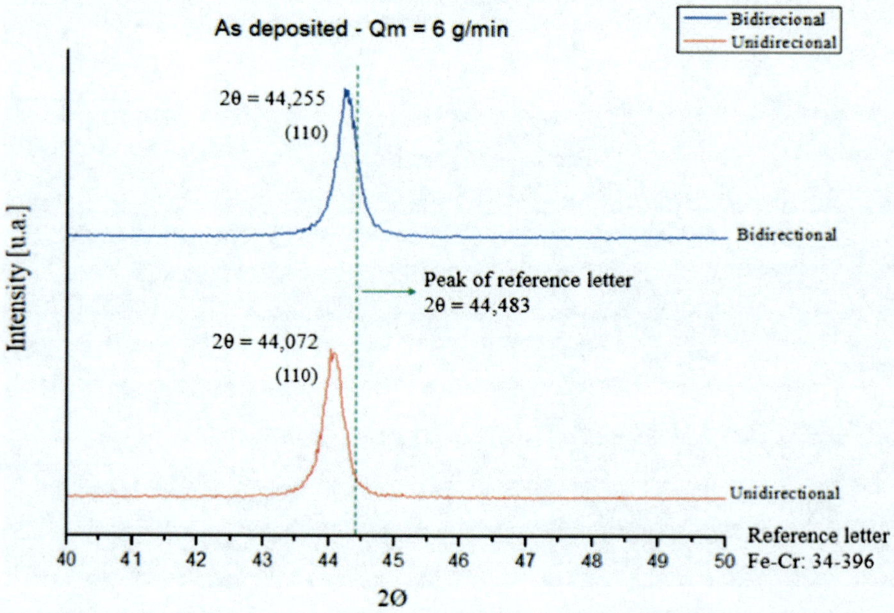

Fig. 8. Detail of the diffractogram peak (110) for last deposited layer in both directions with 6 g/min powder flow rate.

Figure 8 shows the detailed analysis of the X-ray diffractograms in the cross-section of the multilayers deposited with both bidirectional and unidirectional directions in the as-deposited condition with 6 g/min powder flow rate. It shows different displacements of the ferritic phase peaks, indicating differences caused by the deposition direction. Both walls showed an increase in their residual stress, displacement to the right of the peak of the reference chart (34–396), and that the unidirectional deposition direction induces a greater residual stress than the deposition in the bidirectional direction (Turibus 2014). This difference may be related to the thermal cycles associated to each deposition direction. For unidirectional strategy deposition starts when the material is at a lower temperature when compared to that of bidirectional deposition, which facilitates the accommodation of stresses associated with solidification and the contraction and expansion of material already solidified from previous layers.

3.2 Impact of Powder Flow Rate Increasing in the Multilayers

The immediate consequence of the powder flow rate increases is the deposition of thicker layers from 0.64 mm to 0.98 mm, and a smaller number of thermal cycles, for the same processed wall height. Despite the lower number of thermal cycles, the hardness profile remains similar except that the peak hardness at the top region is higher for the higher powder flow rate (9 g/min), Fig. 9.

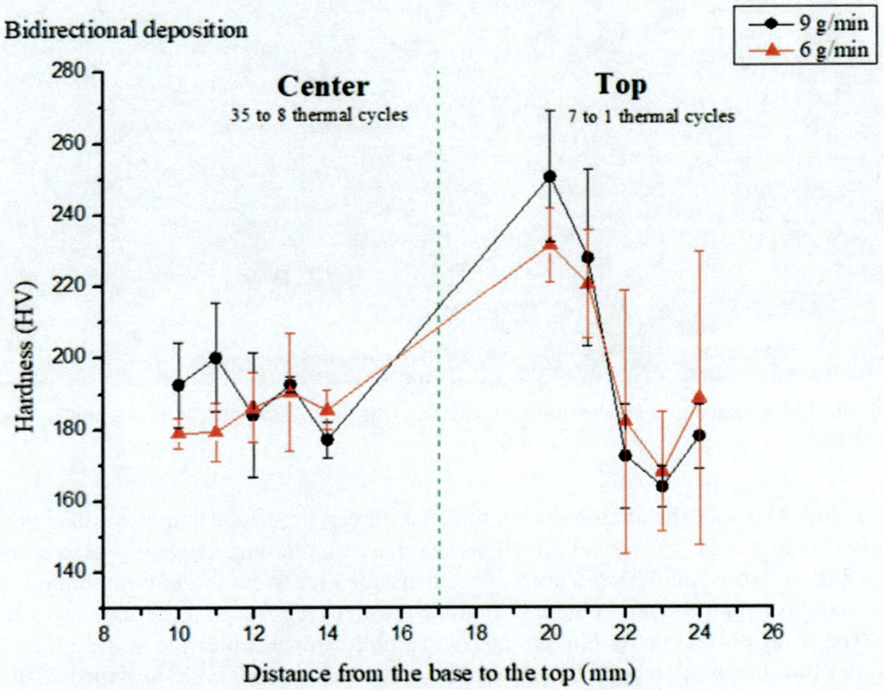

Fig. 9. Hardness profile between the different powder flow rates for the multilayers with the bidirectional deposition.

The temperature to which the material is exposed during multilayer deposition tends to decrease with increasing mass flow rate since more heat from the arc is absorbed to melt the material being deposited, and consequently, the time for air-cooling to the temperature of 150ºC is expected to reduce.

Increasing the feeding rate to 9 g/min, did not change the major features imposed by the deposition strategy as shown in Fig. 5, but resulted in coarser structures, Fig. 10. It is observed that the microstructure of these multilayers presents a similar characteristic to multilayers with 6 g/min, Fig. 6, epitaxial growth, grains with radial orientation in the bidirectional deposition and columnar grains in the unidirectional deposition when analyzed in cross section. It is also noted that, at the top region, the grains are larger when compared to the center region and have an acicular structure along the grain boundaries, identified as Widmanstatten austenite. With the higher feed rate, the material was exposed to a lower temperature and, therefore, it is observed that the primary ferrite grains are coarser when increasing the powder flow rate.

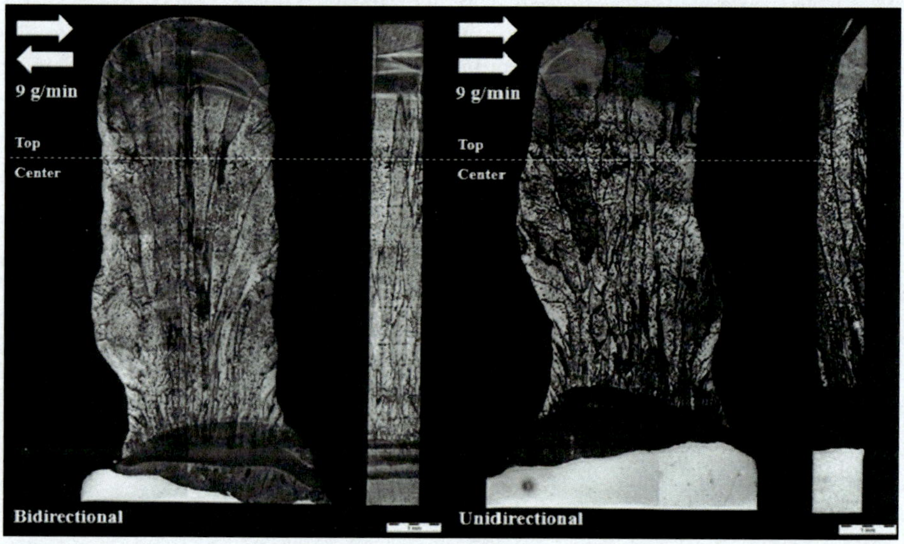

Fig. 10. Bidirectional and Unidirectional walls microstructure deposited with 6 g/min powder flow rate.

Figure 11 details the micrographs for the multilayers processed with 9 g/min in both bidirectional and unidirectional directions that have similar characteristics as seen on Fig. 7. It is also noted the presence of Widmanstatten austenite in the grain boundary and martensite in the central region of multilayers, between 8 and 35 thermal cycles.

The X-ray diffraction for bidirectional and unidirectional deposition to the different powder flow rate used to build the multilayers walls are in the Fig. 12. As expected, the deposition direction and mass flow did not change the ferritic phase formed, as mentioned these peaks can be from either ferrite or martensite. Regardless of the deposition strategy multilayers compared to the smaller feeding rate, Fig. 8, there was a shift to the right and peaks are very similar to those of the reference chart (34–396).

Results show that processing parameters induce changes in microstructure and hardness profiles of multilayers wall processed by PTA-DED that might remain after heat treatments. To mitigate the measured changes thermal management via mapping the

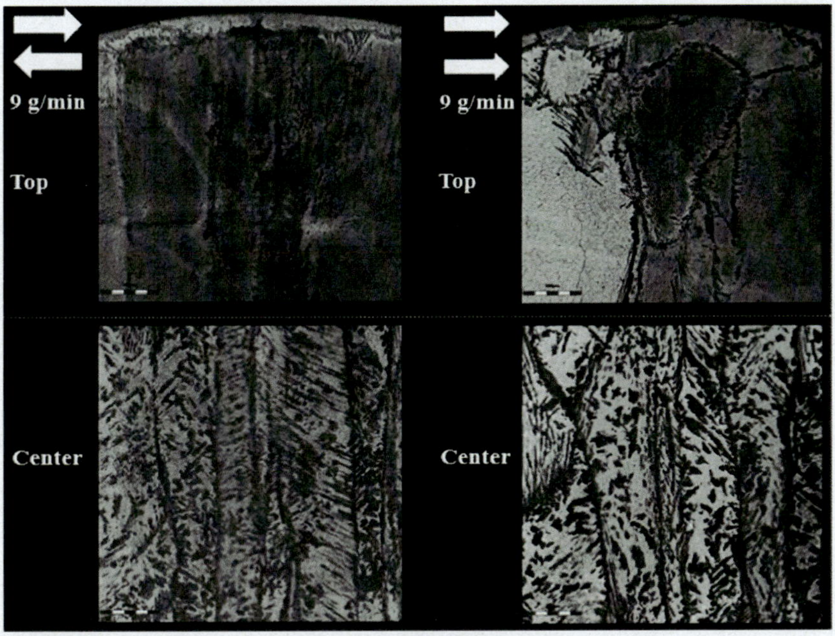

Fig. 11. Cross-section microstructure detail for bidirectional and unidirectional depositions with 9 g/min powder flow rate with 500x enlargement.

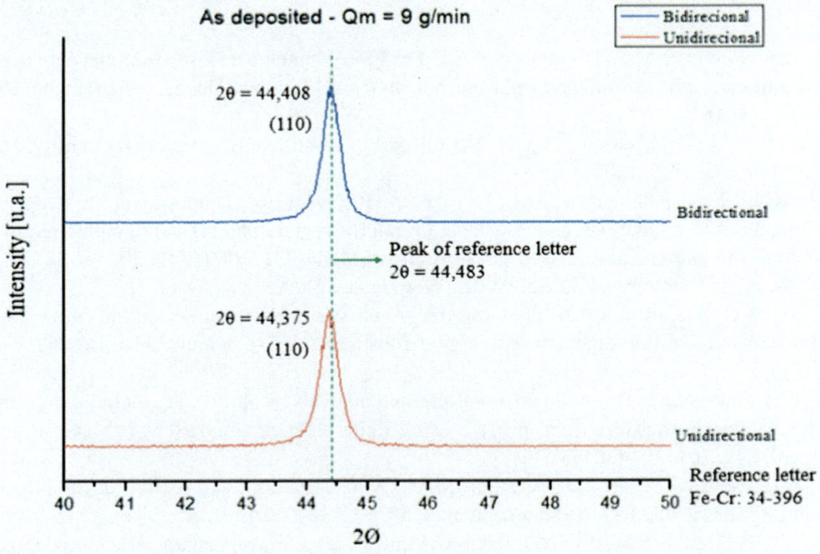

Fig. 12. Detail of the diffractogram peak (110) for last deposited layer in both directions with 9 g/min powder flow rate.

impact of processing parameters can bring important benefits to multilayer processing reducing the need for expensive post heat treatment operations.

4 Conclusions

For the conditions tested in this investigation of the impact of deposition strategy and feeding rate on the features of multilayers of AISI 410L deposited by Plasma transferred arc PTA-DED, it is possible to conclude that:

Deposition strategy induces differences in the microstructure of multilayers characteristics because of different heat flow directions and cooling rates. Epitaxial growth occurs regardless of the processing parameters as confirmed by the long columnar grains at the longitudinal cross section of multilayers. At the transverse cross section of multilayers, bidirectional deposition strategy results in long columnar grains that extend from the bottom to the top layer whereas unidirectional deposition strategy accounts for the smaller grains.

Hardness profile is similar regardless of the processing parameters with a peak hardness near the top and a uniform hardness in the center; the slower cooling rates of unidirectional processing account for the peak hardness occurring closer to the surface.

The increase in the powder flow rate resulted in thicker layers with a lower number of thermal cycles for the same wall height. This also modifies the behavior of the material and its microstructure in the top region, presenting larger and coarser ferrite grains.

References

Alberti, E.A., Bueno, B.M.P., D'Oliveira, A.S.C.M.: Processamento de ligas de níquel com técnica de manufatura aditiva utilizando plasma por arco transferido. Soldagem e Inspeção **20**(2), 137–147 (2015)

Dogu, M.N., et al.: Digitisation of metal AM for part microstructure and property control. Int. J. Mater. Forming **15**(3) (2022)

Khodabakhshi, F., Farshidianfar, M.H., Gerlich, A. P., Nosko, M Trembošová, V., Khajepour, A.: Effects of laser additive manufacturing on microstructure and crystallographic texture of austenitic and martensitic stainless steels. Additive Manuf. **31**, 100915 (2020)

Kou, S.: Welding Metallurgy, 2nd edn. Wiley, New Jersey (2003)

Liu, J., To, A.C.: Deposition path planning-integrated structural topology optimization for 3D additive manufacturing subject to self-support constraint. CAD Comput. Aided Des. **91**, 27–45 (2017)

Nikam, S.H., Jain, N.K.: Three-dimensional thermal analysis of multi-layer metallic deposition by micro-plasma transferred arc process using finite element simulation. J. Mater. Process. Technol. **249**, 264–273 (2017)

Parimi, L.L., Ravi, G., Clark, D., Attallah, M.M.: Microstructural and texture development in direct laser fabricated IN718. Mater. Charact. **89**, 102–111 (2014)

Pereira, H.C.B., Henke, S.L., d'Oliveira, A.S.C.M.: IN625 multilayers characterization deposited by CMT. Soldagem e Inspeção **23**, 235–246 (2018)

Silva, C.C., Machado, J.P.S.E., de Sant, H.B., Farias, J.P.: Estudo da sensitização causada pelo ciclo térmico de soldagem no aço inoxidável superferrítico AISI 444. In: 3° Congresso Brasileiro de P&D em Petróleo e Gás, Salvador (2005)

Thompson, M.K., et al.: Design for additive manufacturing: trends, opportunities, considerations, and constraints. CIRP Ann. Manuf. Technol. **65**(2), 737–760 (2016)

Turibus, S.N.: Análise por difração de raios-X do estado das tensões residuais em chapas de aço inoxidável duplex após soldagem. Universidade Federal do Rio de Janeiro, Rio de Janeiro (2014)

Voort, G.F.V., Lucas, G.M., Manilova, E.P.: Metallography and Microstructures of Stainless Steels and Maraging Steels. ASM International (2004)

Yehorov, Y., da Silva, L.J., Scotti, A.: Exploring the use of switchback for mitigating homoepitaxial unidirectional grain growth and porosity in WAAM of aluminium alloys. Int. J. Adv. Manuf. Technol. **104**, 1581–1592 (2019)

Check for updates

Effects of PTA Deposition Parameters on Geometry and Hardness of AISI 316L Single-Tracks

Gustavo Scheid Prass[1,2](✉), Pedro W. Mattioli[1], and Ana Sofia C. M. d'Oliveira[1]

[1] Federal University of Paraná, Curitiba, PR 81530-000, Brazil
gustavo.prass@ufpr.br

[2] SENAI Innovation Institute in Manufacturing Systems and Laser Processing, Joinville, SC 89218-158, Brazil

Abstract. Directed Energy Deposition (DED) is an Additive Manufacturing (AM) technology involving the layer-by-layer building of components close to their final geometry. One of the main applications of DED is the repair of metallic components, however, geometrical control is challenging due to the nature of feedstock, heat source, and the process itself. Within this context, Plasma Transferred Arc (PTA) is a well-known process for applying coating on metallic materials, guaranteeing a good metallurgical bond between the substrate and the deposited material. This study addresses the effects of deposition current and deposition velocity on the geometry of single-track AISI 316L. The relationship between processing parameters, processability, and hardness is identified and discussed as a valuable database to select AM maintenance procedures. A Design of Experiment (DoE) varying two factors was adopted, deposition current (4 levels) and deposition velocity (3 levels), totaling 12 sets of parameters. Statistical analysis reveals that both factors affect substrate dilution, while a higher current increased dilution the velocity had the opposite effect. The results also showed that both deposition parameters greatly affected the wettability, hence the geometry of the single tracks. The DoE allowed for a good predictive estimate of the interaction with a part being repaired and the refurbishing of its geometry by AM. This research points out that the geometry, microstructure, and hardness of the first deposited track play an important role in the quality and properties of subsequent multilayer builds, required to recover the part geometry or even add functionalities.

Keywords: Additive Manufacturing · Directed Energy Deposition · Plasma Transferred Arc · Design of Experiments

1 Introduction

Additive manufacturing (AM) technology allows components with complex geometry to be manufactured using a layer-by-layer construction. As feedstock, the technology enables the processing of powders or wires from different metallic materials, such as nickel alloys, titanium, aluminum, and stainless steel [Benakis et al., 2020; Karapuzha et al. 2021; Lashgari et al. 2021; Oropeza et al. 2020; Park et al. 2021; Riquelme

D. de Oliveira et al. (Eds.): COBEF 2023, LNME, pp. 140–150, 2024.
https://doi.org/10.1007/978-3-031-43555-3_13

et al. 2021; Wan et al. 2022]. With the capability of processing different metallic alloys, AM can be employed in repairing and remanufacturing damaged parts. This becomes relevant for complex and high-value parts, which would be expensive and time-consuming to be replaced [Priarone et al. 2021]. In refurbishing components, directed energy deposition (DED) is used to add layers of materials to rebuild worn areas of operating parts [Chen et al. 2022; Sahoo and Tripathy, 2020].

DED is an AM technology in which the deposited material is directed by a nozzle and is melted by an energy source, which can be a plasma arc, laser beam, and electron beam [TWI, 2023]. As a general observation, the use of metal powder provides better accuracy and surface finish, but at a lower deposition rate than those used for depositing wire feedstock. Although for both materials (powder and wire) post-processing machining is essential for obtaining the final geometry and surface finish [Priarone et al. 2021]. Notwithstanding the advantages of processing near final geometries without the need for expensive tooling, it is of relevance to control the presence of common defects in AM including crack formation, pores, inclusions, and delamination that strongly depend on the processing parameters must be considered [Li et al. 2021].

Processing parameters also affect the geometrical features of the deposited layer, namely layer height and width, penetration and consequently dilution, in repair operations, these contribute to control the deposit geometry. Furthermore, processing parameters influence the microstructure of the deposited material and have a significant impact on the properties of the material, such as hardness and wear resistance. To make the most of the technology, research has endeavored to optimize the processing parameters for different materials deposited with a variety of energy sources [Bharath et al. 2008; Kumar et al. 2021; Momin et al. 2022; Sadasivam and Amirthalingam, 2022; Wei et al. 2018; Yadav et al. 2019].

In comparison with the laser beam, plasma arc techniques, such as plasma transferred arc (PTA), offer higher deposition rates, good homogeneity, and low oxides content, being a competitive alternative for repairing damaged parts [Cardozo et al., 2019]. However, there is a lack of a systematic investigation on the PTA-DED of AISI 316L that offers information on the processing parameters selection. In particular, for repair operations, the first deposited layer has a two-fold impact on a successful refurbishing of a part: interaction with the original material and it sets a reference for subsequently deposited layers. Thus, in the present study, PTA is used to deposit single tracks of AISI 316L on AISI 304L. A design of experiments (DoE) is carried out to assess the effects of deposition current and speed on the coating hardness and geometry.

2 Materials and Methods

Single layers of gas-atomized AISI 316L stainless steel powder (75–250 μm) were deposited by PTA on AISI 304L plates (70 × 30 × 10 mm). To evaluate the effects of deposition parameters on geometry and hardness, four levels of deposition current and three levels of deposition speed were selected, the processing parameters are shown in Table 1.

Figure 1 shows the geometrical features measured in the cross-section of the single-track, namely height, width, penetration, reinforcement, dilution, and wettability. The hardness and the geometrical features were measured for all experiments in the full factorial design of experiments.

Table 1. PTA deposition parameters.

Parameter	Value
Protective gas	Argon 99%
Carrier gas flow (L/min)	0.8
Plasma gas flow (L/min)	2.0
Shield gas flow (L/min)	15.0
Nozzle-plate distance (mm)	10
Powder flow rate (g/min)	10
Deposition current (A)	60, 90, 120, 150
Deposition speed (mm/min)	100, 150, 200

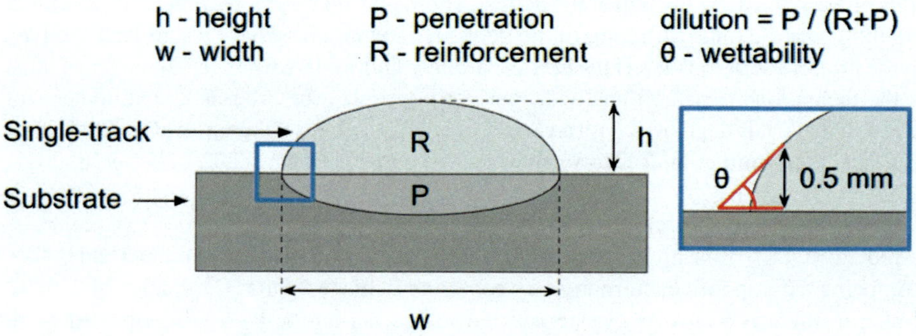

Fig. 1. Single-track geometric features.

3 Results and Discussions

The full factorial experiment was carried out at various combinations of deposition parameters by varying deposition current and speed. Table 2 presents the measured responses of the deposited single-tracks obtained at 12 combinations of parameters, which were repeated in 3 different cross-sections.

Table 2. Experimental conditions and measured responses.

Experiment	Speed (mm/min)	Current (A)	Height (mm)	Width (mm)	Wettability (°)	Reinforcement (mm^2)	Penetration (mm^2)	Dilution (%)	Vickers Hardness (HV 0.3)
1	100	60	2.524	3.712	-	10.254	-	-	169
			2.531	3.704	-	10.169	-	-	174
			2.512	3.735	-	10.310	-	-	178
2	100	90	2.269	5.931	73.11	11.315	0.251	2.17	190
			2.276	5.938	52.80	12.123	0.231	1.87	174
			2.281	5.856	51.00	11.822	0.181	1.51	178
3	100	120	1.863	9.484	36.90	12.460	1.552	11.08	193
			1.856	9.413	40.32	12.422	1.587	11.33	175
			1.863	9.475	42.96	12.679	1.521	10.71	178
4	100	150	1.505	12.036	22.90	12.635	6.053	32.39	189
			1.499	11.972	22.56	12.610	5.904	31.89	189
			1.508	11.999	25.75	12.804	5.998	31.90	179
5	150	60	1.975	3.075	-	5.873	-	-	189
			1.983	3.065	-	5.798	-	-	174
			1.971	3.095	-	5.901	-	-	172
6	150	90	1.724	5.253	58.93	7.161	-	-	186
			1.758	5.039	61.19	7.084	-	-	168
			1.725	5.268	53.18	7.210	-	-	177
7	150	120	1.422	7.631	34.68	8.288	0.815	8.95	184
			1.415	7.591	32.10	8.207	0.756	8.43	183
			1.436	7.583	33.51	8.359	0.776	8.49	177
8	150	150	1.222	9.444	22.43	8.062	4.430	35.46	193
			1.243	9.480	25.30	8.379	4.438	34.63	187
			1.219	9.462	22.32	8.163	4.453	35.30	183
9	200	60	1.596	3.275	-	4.755	-	-	173
			1.587	3.265	-	4.686	-	-	169
			1.599	3.283	-	4.781	-	-	177
10	200	90	1.520	4.862	39.73	5.913	-	-	189
			1.532	4.450	46.99	5.601	-	-	173
			1.532	4.450	46.99	5.601	-	-	173
11	200	120	1.183	6.663	33.18	5.918	0.800	11.91	178
			1.169	6.649	27.11	5.867	0.805	12.07	181
			1.160	6.707	25.95	5.917	0.768	11.49	183
12	200	150	1.017	8.448	17.81	6.087	3.695	37.77	196
			1.036	8.485	18.82	6.369	3.388	34.72	180
			1.025	8.476	17.83	6.227	3.316	34.75	183

The optical microscopies of the single-track cross-sections are shown in Fig. 2. It reveals a lack of fusion for the tracks deposited at a lower current (60 A) that are detached from the substrate. Lack of fusion was also identified at the following deposition with 90 A, and the faster speeds tested, 150 and 200 mm/min, as a consequence of insufficient energy per unit length. The cross-sections of the processed layers reveal that both current and speed have a great impact on the geometry of the deposited single-tracks.

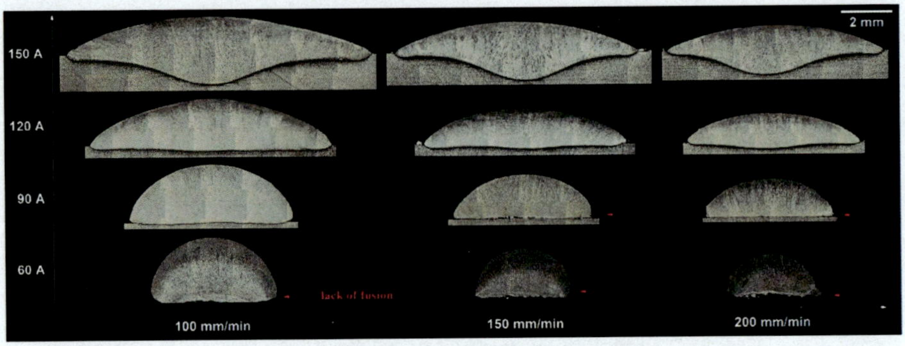

Fig. 2. Optical microscopies of the single-track cross-sections.

Process maps of deposition current and speed, Fig. 3 shows the contour plots of the measured geometrical features and hardness. When deposition speed is increased, less powder is fed into the melt pool per length unit, since the powder flow rate (10 g/min) is kept constant. As a result, higher deposition speeds lead to lower track heights (Fig. 3a) and lower track widths (Fig. 3b). As expected, a decrease in reinforcement area (Fig. 3d) is observed with an increase in deposition speed. Furthermore, higher deposition speed causes a reduction in energy per unit length, slightly decreasing the penetration area measured (Fig. 3e). For the tested conditions, dilution (Fig. 3f) remains almost constant with changes in speed, since both reinforcement and penetration decrease with higher deposition speed [Santos et al. 2019].

Alongside speed, deposition current also highly impacts the single-track geometry. When the deposition current is increased, more energy is available per unit length, leading to higher temperatures in the melt pool. As a consequence, higher deposition currents lead to lower wettability angles (Fig. 3c), decreasing track height (Fig. 3a), and substantially increasing the measured track width (Fig. 3b). Although keeping the powder flow rate constant, increasing deposition current slightly increases reinforcement area (Fig. 3d), this can be explained by the better capturing of the metallic powder, since higher temperatures lead to a wider melt pool (Fig. 2). Higher currents lead to an increase in both penetration and dilution (Fig. 3e and Fig. 3f) since more energy is available at the melt pool [Yaedu and d'Oliveira, 2005; Takano et al., 2008].

Analysis of variance (ANOVA) was used to quantify the impact of both individual and combined deposition parameters. Tables 3–9 show the ANOVA data for the measured responses. At Table 3, it is seen that both speed (46.66%) and current (50.65%) significantly contributed to the track height response. Whereas track width (Table 4) and wettability (Table 5) were mainly influenced by deposition current (86.73% and 91.92%, respectively), which highlights the effect of the energy input on these variables. The fact that the powder flow rate per length unit depends on the speed is emphasized in Table 6, where reinforcement is highly impacted by deposition speed (89.94%). In turn, penetration (Table 7) and dilution (Table 8) were shown to be dependent on heat input, since deposition current has the maximum contribution (92.08% and 99.22%, respectively).

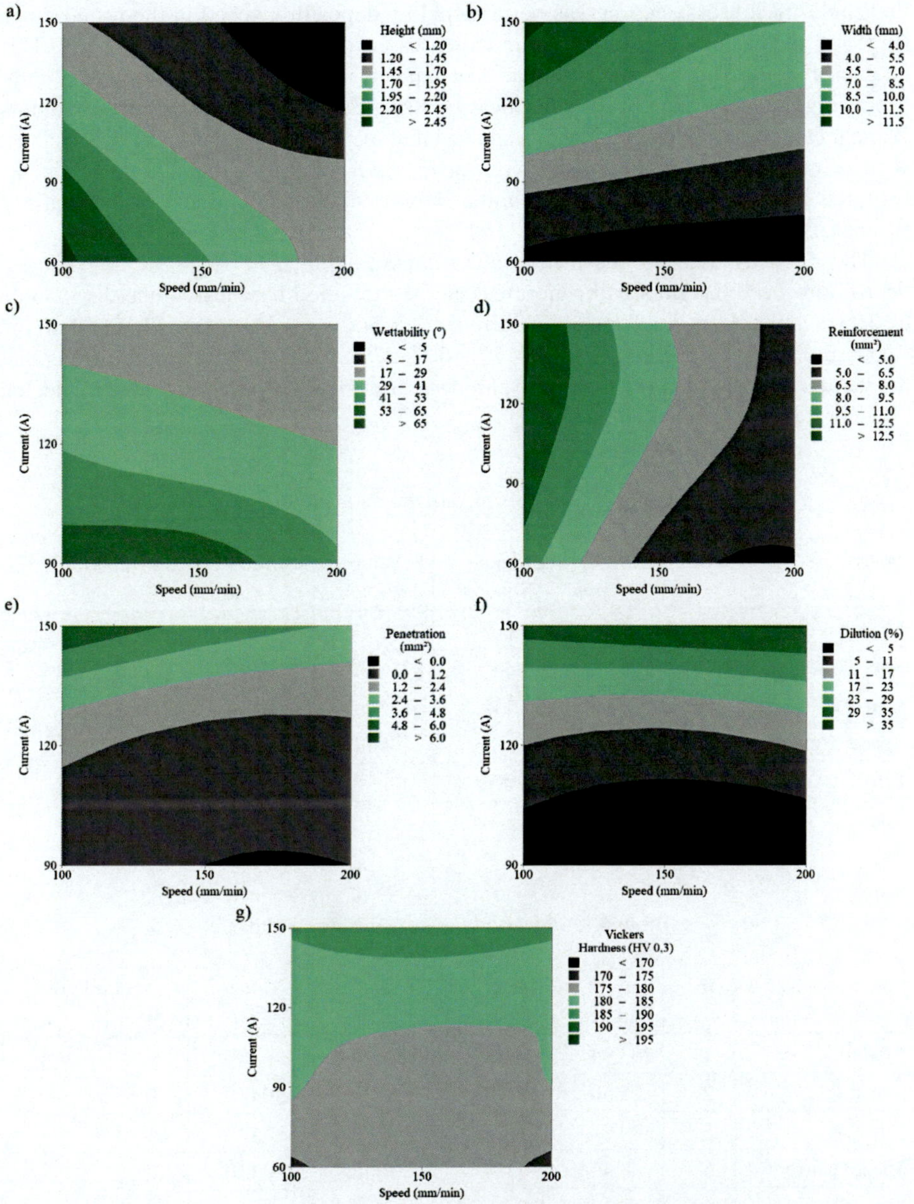

Fig. 3. Contour plot of (a) Height, (b) Width, (c) Wettability, (d) Reinforcement, (e) Penetration, (f) Dilution, and (g) Vickers Hardness.

Since the feedstock and the substrate have different chemical compositions, the coating hardness is directly impacted by dilution [Takano et al. 2008]. As aforementioned, dilution was not impacted by deposition speed, therefore hardness measured at the single-track cross-section was not affected by deposition speed in the tested range (Fig. 3g). Increasing deposition current increases the measured hardness (Fig. 3g). This is a surprising result because a higher heat input leads to a slower cooling rate compromising the refinement of the solidification structure accounting for a lower hardness [Kumar et al. 2014]. However, results showed that the increase in hardness is associated with an increase in dilution with the substrate material suggesting that alloying elements from the substrate increased solid solution saturation and the hardness of the coatings [Ramakrishnan et al. 2021].

The ANOVA data for the measured hardness is shown in Table 9. As expected, deposition speed has no significant effect on the measured hardness, since the P-value of $0.93 > 0.05$. Also, the deposition current contribution on hardness is lower than the standard deviation contribution (33.24% and 62.37%, respectively), which means that the standard deviation of the measured hardness surpassed the variation with the selected range of deposition parameters tested.

Table 3. Analysis of variance data for height.

Source	DF	Sum of Squares	Mean Square	F-Value	P-Value	Contribution %
Model	11	6.58	0.60	3419.47	< 0.001	-
Speed	2	3.07	1.54	8780.89	< 0.001	**46.66**
Current	3	3.34	1.11	6354.55	< 0.001	**50.65**
Speed*Current	6	0.17	0.03	164.80	< 0.001	2.63
Error	24	0.00	0.00			0.06
Total	35	6.59				100.00

Table 4. Analysis of variance data for width.

Source	DF	Sum of Squares	Mean Square	F-Value	P-Value	Contribution %
Model	11	263.95	24.00	4256.28	< 0.001	-
Speed	2	25.50	12.75	2261.09	< 0.001	9.65
Current	3	229.04	76.35	13542.03	< 0.001	**86.73**
Speed*Current	6	9.42	1.57	278.47	< 0.001	3.57
Error	24	0.14	0.01			0.05
Total	35	264.09				100.00

Table 5. Analysis of variance data for wettability.

Source	DF	Sum of Squares	Mean Square	F-Value	P-Value	Contribution %
Model	11	14151.30	1286.48	71.89	< 0.001	-
Speed	2	448.90	224.47	12.54	< 0.001	3.08
Current	3	13402.80	4467.60	249.65	< 0.001	**91.92**
Speed*Current	6	299.60	49.93	2.79	0.033	2.05
Error	24	429.50	17.90			2.95
Total	35	14580.80				100.00

Table 6. Analysis of variance data for reinforcement.

Source	DF	Sum of Squares	Mean Square	F-Value	P-Value	Contribution %
Model	11	267.97	24.36	1007.46	< 0.001	-
Speed	2	241.52	120.76	4994.19	< 0.001	**89.94**
Current	3	24.66	8.22	340.00	< 0.001	9.18
Speed*Current	6	1.78	0.30	12.27	< 0.001	0.66
Error	24	0.58	0.02			0.22
Total	35	268.55				100.00

Table 7. Analysis of variance data for penetration.

Source	DF	Sum of Squares	Mean Square	F-Value	P-Value	Contribution %
Model	11	139.46	12.68	3041.51	< 0.001	-
Speed	2	4.91	2.45	588.69	< 0.001	3.52
Current	3	128.51	42.83	10276.40	< 0.001	**92.08**
Speed*Current	6	6.04	1.01	241.67	< 0.001	4.33
Error	24	0.10	0.00			0.07
Total	35	139.56				100.00

Table 8. Analysis of variance data for dilution.

Source	DF	Sum of Squares	Mean Square	F-Value	P-Value	Contribution %
Model	11	6995.12	635.92	2049.48	< 0.001	-
Speed	2	5.72	2.86	9.21	< 0.001	0.08
Current	3	6948.18	2316.06	7464.34	< 0.001	**99.22**
Speed*Current	6	41.22	6.87	22.14	< 0.001	0.59
Error	24	7.45	0.31			0.11
Total	35	7002.57				100.00

Table 9. Analysis of variance data for hardness.

Source	DF	Sum of Squares	Mean Square	F-Value	P-Value	Contribution %
Model	11	705.64	64.15	1.32	0.275	-
Speed	2	7.06	3.53	0.07	0.93	0.38
Current	3	623.19	207.73	4.26	0.015	33.24
Speed*Current	6	75.39	12.57	0.26	0.951	4.02
Error	24	1169.33	48.72			**62.37**
Total	35	1874.97				100.00

4 Conclusions

A systematic investigation on the PTA-DED of AISI 316L was carried out in this study. The key findings for the tested conditions are as follows:

- AISI 316L powder was successfully deposited by PTA, resulting in sound single-track coatings. The exception was the coatings processed with 60 A, which resulted in insufficient dilution, detaching the coating from the substrate.
- The contour plots can be used to predict the hardness and the geometric response of AISI 316L coating to process parameters within the tested range.
- Single track height is governed by both deposition current and speed, which decreases by increasing any of the two tested factors.
- Deposition current was found to be the most influential parameter affecting width, as a result of a decrease in wettability angle. In addition, penetration, and consequently dilution, is mostly influenced by deposition current.
- On the other hand, deposition speed was discovered to be the most significant factor influencing reinforcement area, since the powder flow rate was kept constant.
- For the tested conditions, the variation in deposition speed did not change the hardness of the AISI 316L coating. However, the dispersion of the measured hardness is very significant for the test conditions used.

Acknowledgements. The authors acknowledge CAPES, CNPq and the Laboratory of Additive Manufacturing and Surface Engineering (LAMSE) of UFPR.

References

Benakis, M., Costanzo, D., Patran, A.: Current mode effects on weld bead geometry and heat affected zone in pulsed wire arc additive manufacturing of Ti-6-4 and Inconel 718. J. Manuf. Process. **60**(18), 61–74 (2020)

Bharath, R.R., Ramanathan, R., Sundararajan, B., Srinivasan, P.B.: Optimization of process parameters for deposition of Stellite on X45CrSi93 steel by plasma transferred arc technique. Mater. Des. **29**(9), 1725–1731 (2008)

Cardoso, E., Rios, S., Pardal, G., Ganguly, S., D'Oliveira, A.S.C.M.: Additive techniques to refurbish ni based components. Soldagem e Inspeção **24**, 8 (2019)

Chen, X., Wang, C., Ding, J., Bridgeman, P., Williams, S.: A three-dimensional wire-feeding model for heat and metal transfer, fluid flow, and bead shape in wire plasma arc additive manufacturing. J. Manuf. Process. **83**, 300–312 (2022)

Karapuzha, A.S., Fraser, D., Zhu, Y., Wu, X., Huang, A.: Effect of solution heat treatment and hot isostatic pressing on the microstructure and mechanical properties of Hastelloy X manufactured by electron beam powder bed fusion. J. Mater. Sci. Technol. **98**(6), 99–117 (2021)

Kumar, R., Arya, H.K., Saxena, R.K.: Experimental determination of cooling rate and its effect on microhardness in submerged Arc welding of mild steel plate (Grade c-25 as per IS 1570). J. Mater. Sci. Eng. **3**(2), 1000138 (2014)

Kumar, P., Sawant, M.S., Jain, N.K.: Optimization of process parameters in micro-plasma transferred arc deposition process for cobalt-based alloy. Mater. Today: Proc. **44**, 1681–1686 (2021)

Lashgari, H.R., Li, S., Kong, C., Asnavandi, M., Zangeneh, S.H.: Rotary friction welding of additively manufactured 17–4PH stainless steel. J. Manuf. Process. **64**(16), 1517–1528 (2021)

Li, R., Jin, M., Paquit, V.C.: Geometrical defect detection for additive manufacturing with machine learning models. Mater. Des. **206**(20), 109726 (2021)

Momin, A.G., Khatri, B.C., Chaudhari, M., Shah, U., Valaki, J.: Parameters for cladding using plasma transfer arc welding – A critical. Mater. Today: Proc. **77**(3), 614–618 (2022)

Oropeza, D., et al.: Welding and additive manufacturing with nanoparticle-enhanced aluminum 7075 wire. J. Alloy. Compd. **834**(13), 154987 (2020)

Park, S.H., Son, S.J., Lee, S.B., Yu, J.H., Ahn, S.J., Choi, Y.S.: Surface machining effect on material behavior of additive manufactured SUS 316L. J. Market. Res. **13**(22), 38–47 (2021)

Priarone, P.C., Campatelli, G., Catalano, A.R., Baffa, F.: Life-cycle energy and carbon saving potential of Wire Arc Additive Manufacturing for the repair of mold inserts. CIRP J. Manuf. Sci. Technol. **35**, 943–958 (2021)

Ramakrishnan, A., Rameshkumar, T., Rajamurugan, G., Sundarraju, G., Selvamuthukumaran, D.: Experimental investigation on mechanical properties of TIG welded dissimilar AISI 304 and AISI 316 stainless steel using 308 filler rod. Mater. Today: Proc. **45**, 8207–8211 (2021)

Riquelme, A., Candela, C.S.R., Ram, J.: Influence of process parameters in additive manufacturing of highly reinforced 316L / SiCp composites. J. Manuf. Process. **299**(14), 117325 (2021)

Sadasivam, P., Amirthalingam, M.: Design and fabrication of micro-plasma transferred wire arc additive manufacturing system. CIRP J. Manuf. Sci. Technol. **37**, 185–195 (2022)

Sahoo, A., Tripathy, S.: Development in plasma arc welding process: A review. Mater. Today: Proc. **41**, 363–368 (2021)

Santos, A.X., et al.: Study on influence of the PTA-P welding process parameters on corrosion behavior of Inconel 625 coatings. Matéria (Rio J.) **24**(1) (2019)

Takano, E.H., Queiroz, D., d'Oliveira, A.S.C.M.: Avaliação dos Parâmetros de Pro-cessamento por PTA nas Superfícies Processadas. Soldagem & Inspeção **13**, 210–218 (2008)

TWI: What is directed energy deposition (DED). https://www.twi-global.com/technical-knowledge/faqs/directed-energy-deposition, (Accessed 21 May 2023)

Wan, H.Y., et al.: Toward qualification of additively manufactured metal parts: Tensile and fatigue properties of selective laser melted Inconel 718 evaluated using miniature specimens. J. Mater. Sci. Technol. **97**(29), 239–253 (2022)

Wei, Y., Wei, X., Chen, B., Zuo, J., Ma, T., Shen, J.: Parameter optimization for tungsten carbide/Ni-based composite coating deposited by plasma transferred arc hardfacing. Trans. Nonferrous Metals Soc. China **28**(12), 2511–2519 (2018)

Yadav, S., Jinoop, A.N., Sinha, N., Paul, C.P., Bindra, K.S.: Parametric investigation and characterization of laser directed energy deposited copper-nickel graded layers. Inter. J. Advanced Manuf. Technol. **108**(11–12), 3779–3791 (2020)

Yaedu, A.E., d'Oliveira, A.S.C.M.: Cobalt based alloy PTA hardfacing on different substrate steels. Mater. Sci. Technol. **21**(4), 459–466 (2005)

Effect of Laser Power and Powder Morphology on Surface Roughness of TI6Al4V Produced by Laser Powder-Directed Energy Deposition

Geovana Eloizi Ribeiro[1](✉), Vincent Wong[1], Willian Roberto Valiceli Sanitá[1], Alessandro Roger Rodrigues[1], and Reginaldo Teixeira Coelho[2]

[1] Mechanical Engineering Department, São Carlos School of Engineering, University of São Paulo, São Carlos, Brazil
geovana_rib@usp.br

[2] Production Engineering Department, São Carlos School of Engineering, University of São Paulo, São Carlos, Brazil

Abstract. Additive manufacturing of metals has emerged as a technology capable of producing complex metal parts in the "near net shape" format, performing repairs, and creating components with gradient material, enabling manufacturing parts with high added value and low production. Directed Energy Deposition from Laser and Powder (LP-DED) is one of the categories of the additive manufacturing process by which concentrated thermal energy allows the metallic powder to melt. These applications have been attractive to different areas such as aerospace, automotive, and medical. In the medical field, its application has focused on creating implants, prostheses, instruments, and medical devices. Ti6Al4V titanium alloys have stood out due to their high mechanical strength, high corrosion resistance, low density, and good biocompatibility. One of the literature challenges reflects the roughness given to printed parts by the LP-DED process, which can affect the osseointegration of prostheses and implants, linked to their recovery time and success. This article evaluates the roughness of Ti6Al4V parts obtained from the LP-DED process using two types of powder. The first is produced by gas atomization, and the second by advanced plasma atomization. Subsequently, eight specimens were fabricated by LP-DED on pure Ti substrate. The laser power was another input variable ranging from 300 W to 345 W with a 15 W increment. The samples were cleaned with deionized water and acetone using ultrasonic vibration. Then, we evaluated the roughness of the samples using a confocal microscope. The roughness evaluation shows that a particle size distribution with Gaussian behavior, as demonstrated by AP&C, resulted in a coarser roughness. In contrast, the Carpenter Additive powder resulted in a slightly thinner roughness. The Laser Power influenced the surface quality due to the increment of density energy. The results obtained in this article represent a breakthrough in medical implants, offering solutions that enable surface integrity and osseointegration, improving the rehabilitation process and increasing the quality.

Keywords: Directed Energy Deposition · Roughness · Ti6Al4V · Additive Manufacturing

D. de Oliveira et al. (Eds.): COBEF 2023, LNME, pp. 151–160, 2024.
https://doi.org/10.1007/978-3-031-43555-3_14

1 Introduction

Titanium alloy is a material that possesses high specific strength, moderate Young's Modulus, and excellent corrosion resistance; these attributes have been used in areas such as aerospace, marine, chemical, military, and biomedical [1]. In biomedical applications, titanium alloy is broadly used in implants and prostheses. These devices spend most of their cycle of life inside the human body. For this reason, titanium alloy needs to satisfy some requirements related to compatibility (Mechanical, Chemical, and Biological) between the user and the material. When the device's compatibility is improved, it also enhances the function of the implant, which works similarly to the original tissue. It reduces the risk of suffering widely known diseases that are mostly associated with poor functionality of the implant. Ti6Al4V alloy is an $\alpha + \beta$ alloy, considered a first-generation alloy most used. The Al is added to stabilize the α phase, which increases the β transformation temperature, while V stabilizes the β phase and reduces it.

Additive Manufacturing (AM) is a set of techniques that enable the creation of metallic parts with an agreed value high and a low production level. This technique is based on adding material layer upon layer, contrary to subtractive manufacturing [2]. It has been used in many materials, including metals, polymers, ceramics, and composites [3]. There are three criteria to classify Metal Additive Manufacturing (MAM): How the material is prepared, the type of material used, and the type of energy source. From the first criteria, two groups can be found: Powder Bed Fusion (PBF) and Directed Energy Deposition (DED). In DED, the type of material can be powder and wire, and the heat source can be from Laser, Plasma Arc, and Electron Beam melting. Specifically, Laser Powder Directed Energy Deposition (LP-DED) uses the incidence of a high-power laser on metallic powder, coaxially delivered, forming a melt pool that solidifies on a substrate.

Although MAM has numerous advantages, surface finishing of fabricated parts remains one of the major concerns [4]. In biocompatibility terms, the surface roughness plays a role important in guaranteeing osseointegration [5]; recently, studies have shown the influence of the process parameters such as deposition strategy, process parameters, layer thickness, laser power, and scanning speed in the part roughness [6–8]. However, the results observed in the literature using LP-DED showed a different range of surface roughness due to the powder's quality, including (morphology, particle size), machine configuration, the necessity of a controlled atmosphere, and adjusting the process parameters.

In the biomedical area, research indicates that implant fixation is a product of surface roughness and bone position [9]. In roughness terms, the surfaces generated by the LP-DED method are characterized by rough aspects [10]. In shear cases, surfaces with higher roughness offer greater bonding with bone tissue, providing greater resistance to ruptures [11].

According to Ye, surface finishing plays an important role in the fatigue performance of the part [12]. Generally, the parts produced by MAM need to present a better surface finish in as-built conditions [13, 14]. It is due to three reasons: 1) The adherence of semi-melted powder on the part's surface. This phenomenon is called balling and results from regrouping powder particles under inhomogeneous heating conditions [15]; 2) The staircase phenomenon. It allows visualizing the layer mark caused by the offset in part with an inclined or curved surface [16]; 3) and the appearance of pores or regions with incomplete melting. Greitemeier highlighted that although the mechanical properties, microstructure, and other factors affect the fatigue performance in parts produced by MAM, surface finishing is the main cause of poor fatigue performance compared with milled parts [17].

Chan analyzed the fatigue life of titanium alloys manufactured by Electron Beam Melting and Laser Beam Melting. They established a double logarithm plot of various Ti6Al4V samples that allow the correlation of the maximum surface roughness (Ra) with the mean fatigue life cycle. Also, the Laser Beam melting samples obtained a longer life than Electron Beam Melting [18].

Various roughness ranges are suitable for biomedical implants; the values vary according to the raw material, type of manufacturing, desired application, and post-processing. According to Svetlizky, good dental implants should have Ra between 0.5 and 1.0 μm, and it is important to consider that the removal torque increases when growing Ra up to a certain value [19]. Wennerberg and Albrektsson found commercial Sa roughness ranging from 0.54 to 2.09 μm [20]. However, larger Ra roughness intervals between 45 and 53.25 μm showed a high cell proliferation and adhesion capacity in Titanium [21]. In addition, according to Hatamleh, Sq roughness from 2.81 to 16.68 μm indicated equal biocompatibility when considering different surface treatments in Ti6Al4V implants such as polishing, sandblasting, acid etching, and electrochemical treatment [22]. Finally, Krishna declared that bone tissue could adapt to surface irregularities with 1 to 100 μm height and that changing the surface topography of an implant can greatly improve its stability [23].

This paper aims to establish the relationship between part roughness, powder morphology, and laser power when printing Ti6Al4V by LP-DED and compare these results to those found in the literature, indicating some possible applications.

2 Materials and Methods

The LP-DED machine BeAM Modulo 250 and Ti6Al4V powder (grade 23) manufactured by Carpenter Additive and Ti6Al4V powder (grade 5) manufactured by AP&C were applied in this study. The chemical composition of each powder is described in Table 1. [24, 25]. The Argon was a carrier, central, and shield work gas. The scanning speed of 2000 mm/min and the powder feed rate of 5.4 g/min were kept constant. The material used as a substrate was composed of pure titanium close to the deposition area of the cubes. The dimensions of the specimens fabricated were 15 mm in length, 15 mm in width, and 6 mm in height. The process parameters of each sample are described in Table 2.

Before measuring the roughness, the samples were cleaned with acetone and deionized water under ultrasonic vibration at 35 kHz by employing a sonicator machine, Julabo USR 1, to remove the powder particles that adhered to the study surface. Besides average center roughness (Sa) typically utilized for biomedical applications, this research also measured the Root Mean Square Height (Sq), skewness (Ssk), and kurtosis (Sku) roughness by using a microscope confocal Olympus OLS4100.

Table 1. Chemical composition of both study powders, including the variation of O and N as an interstitial element.

	Al	C	H	Fe	N	O	Ti	V	Y
Carpenter	5.5–6.5	0–0.8	0–0.01	0–0.25	0–0.03	0–0.13	balance	3.5–4.5	0–0.05
AP&C	5.5–6.75	0–0.8	0–0.01	0–0.4	0–0.05	0–0.2	balance	3.5–4.5	–

Table 2. Process Parameters used with laser power between 300 W to 345 W.

Specimen #	Laser Power (W)	Powder manufacturer	Specimen #	Laser Power (W)	Powder manufacturer
1	300	Carpenter	1	300	AP&C
2	315		2	315	
3	330		3	330	
4	345		4	345	

3 Results and Discussions

3.1 Powder Characterization

As mentioned above, two powders were used in this study. This powder has mostly spherical morphology with the presence of satellites, pores, and flakes to a lesser extent. The first powder was a Carpenter Ti6Al4V ELI grade 23 alloy produced by gas atomization and shown in Fig. 1a. The particle size ranges from 45 - 106 μm. The results of the particle size distribution of the powders are shown in Fig. 1b, which was measured by the optical microscope Olympus OLS4100.

The second Powder, Ti6Al4V grade 5 alloy produced by advanced plasma atomization (APATM) and shown in Fig. 1c, was provided by AP&C. The particle size ranges from 45–90 μm. This powder has mostly spherical morphology, with less presence of flakes and satellites when compared with Carpenter powder and particle size close to the mean. The results of the particle size distribution of the powders are shown in Fig. 1d. The main difference between both types of powder is the reduction of the content of interstitial elements such as N or O. This difference plays a key role in microstructures because they are strong α-stabilizers and influence the α-to-β transition temperature. Low interstitial element content improves ductility and fracture toughness [26]. In Ti6Al4V, Grade 23, the reduction of oxygen content is limited to 0.13% (maximum). In Ti64Al4V grade 5, the reduction of oxygen content is restricted to 0.2% (max).

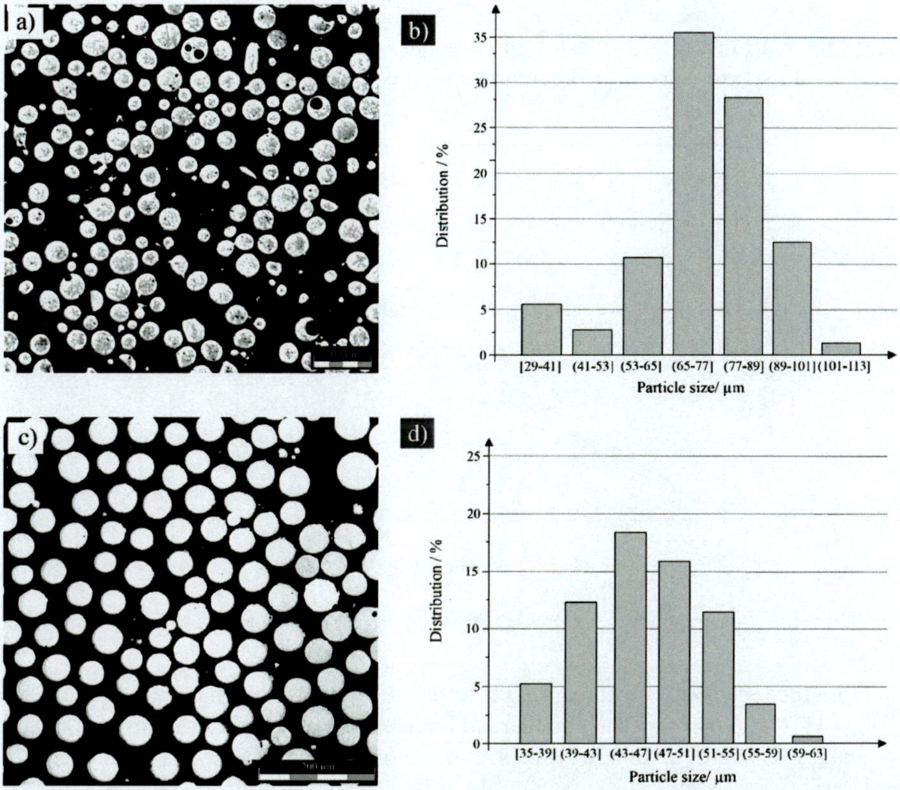

Fig. 1. Optical microscopy powder a) Carpenter with the presence of satellite, pores, and flakes to a lesser extent. b) Particle size distribution Ti6Al4V Carpenter powder with non-gaussian behavior. c) AP&C with particle size close to the mean. d) Particle size distribution Ti6Al4V AP&C powder with behavior close to the Gaussian distribution.

3.2 Process Parameters

In this study, the powder feed rate and laser powder were essential to construct cubes with uniform geometry. The powder feed rate was obtained by capturing and measuring the powder mass for one minute using a horizontal disk powder feeder. To Carpenter powder, the rotation of the horizontal feed disk was defined in values of 1, 1.4, and 2 rpm. To AP&C powder, the rotation was determined at 1, 1.4, and 1.6 rpm. This procedure was performed three times for each rotation to ensure measurement reliability.

When combined with the laser power, the powder particle size and the powder feed rate directly affect the porosity and tiny void generation in LP-DED, as described by Averardi [27]. In Fig. 2, it is possible to observe that the precision of the powder feed rate is more efficient when the particle size has a behavior similar to the Gaussian distribution. In terms of laser power, all the samples showed a good grip, without burns, and with a linear deposition. At the beginning of the deposition process, the melt pool was composed predominantly of powder, creating a small lack of fusion at the beginning of the layers. Using the substrate near the deposition area allowed the creation of density energy focused, favoring the fusion in subsequent layers, creating a good underlayer adhesion of the laser spot at the beginning and the final of the deposition trajectory.

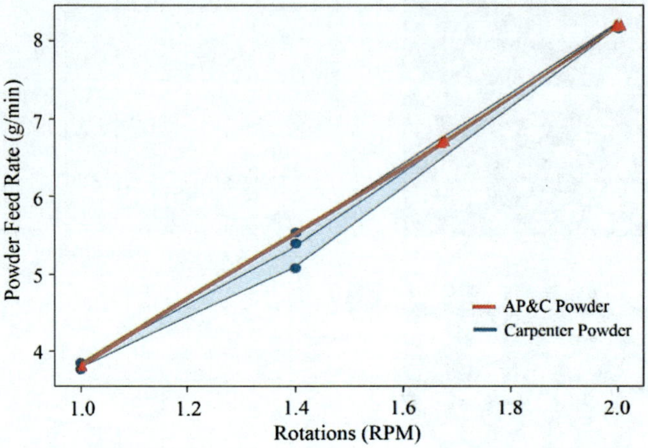

Fig. 2. Rotation (rpm) vs. powder feed rate to both powders.

From the viewpoint of geometrical features, the dimensions obtained of parts were 15.73 x 15.77 x 6.04 mm. The geometrical features of one sample are shown in Fig. 3. The values in X and Y superior of 15 mm are associated with the spot diameter of the laser. When the deposition occurs, the origin of the coordinate does not consider the middle of the circumference of the laser spot at the beginning and the final of the deposition trajectory.

3.3 Roughness Evaluation

The finishing data revealed a tendency in both averages (Sa) and quadratic (Sq) roughness parameters to decrease when increasing the laser power, as seen in Fig. 4. The powder is more densely melted since the degree of fusion increases with laser power. Furthermore, comparing the two powders of different manufacturers, the AP&C powder showed roughness ranges with higher values relative to the Carpenter one.

These Sa roughness findings are slightly lower than those of Zareidoost [21], while the Sq ones are higher than those of Hatamleh [22]. Such differences result from the as-built condition of the samples studied in this paper and the post-processing or surface treatment carried out in specimens from the literature. A way to reach the desired roughness ranges, such as those obtained by Wennerberg and Albrektsson, is to change the process parameters once the Sa and Sq roughness decrease as the laser power increases, allowing receiving surfaces with better finishing and less need for post-processing [20].

Examining the statistical parameters skewness and kurtosis from Fig. 5, it is noticed that Ssk values are positive regardless of the laser power and powder provider, which means that the surface produced by LP-DED presents material concentration on valleys.

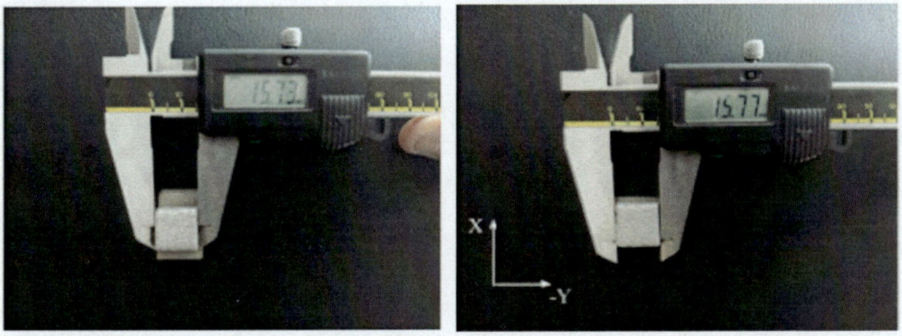

Fig. 3. Geometrical features of one sample show a measure superior to the initial geometry.

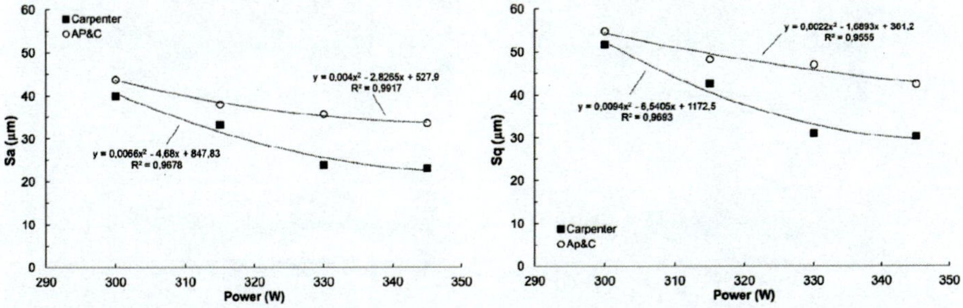

Fig. 4. Sa and Sq roughness as a function of laser power and Ti powder provider.

Negative levels would indicate a plateau-based surface or flat peaks. Although the skewness analysis can be ternary (negative, zero, or positive), two minimum points for each powder manufacturer are identified in 321 W and 340 W, respectively, for Carpenter and AP&C providers. This finding can be particularly useful since positive skewness magnitudes were measured by showing a surface highly skewed with a greater absence of material on the sample surface. When evaluating the kurtosis, it is verified that all values are greater than three, which means that the printed surface presents large peaks or valleys (spiky surface). This surface texture produced by LP-DED can be interesting when applied to implants, i.e., a spiky topography (Sku > 0) with a predominance of peaks (Ssk > 0) could work as cells housing, maximize their attachment at microscope level, and consequently, the formation of bone matrix around the implant.

In addition to the mean and profile quantitative roughness analysis, a qualitative evaluation of the printed surfaces is also investigated. Figure 6 shows the samples' topography by color 3D graphs by which it is possible to verify the symmetry of the peaks (red) and valleys (blue) and the distribution of the pronounced peaks or valleys.

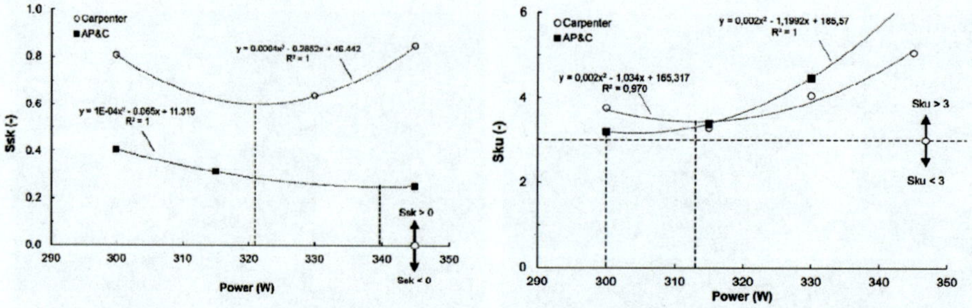

Fig. 5. Ssk and Sku roughness as a function of laser power and Ti powder provider.

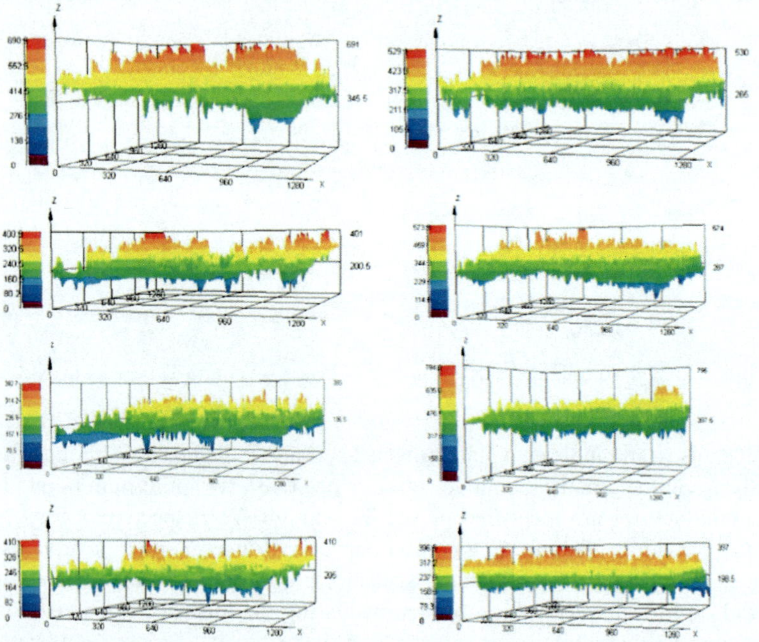

Fig. 6. 3D surfaces obtained from LP-DED of Ti6Al4V as a function of powder provider and laser power.

4 Conclusions

This paper determined a relationship between surface roughness (CLA, RMS, skewness, and kurtosis) generated by additive manufacturing through the LP-DED process of Ti6Al4V powders for biomedical applications and process parameters (laser power and powder provider). The following conclusions can be drawn based on the results.

- **The particle size influences the surface roughness due to small particles, as observed** by the Carpenter powder not governed by gravity. Inter-particle forces such as Van der walls and electrostatic forces become more important. Also, the balling phenomenon favors the size of smaller particles, adhering finer particles to the part's surface. This research showed that the laser power and the powder morphology play an important role in the roughness of the region, improving the quality of devices for medical applications.

- The laser power also contributes to the decrease of the roughness. It is due to the laser power being directly correlated to global density energy and the powder capture efficiency. A large melt pool represents high global density energy, and consequently, a large melt pool contributes to the major high powder capture efficiency. Average roughness Sa decreased by about 42% and 23%, respectively, for Carpenter and AP&C powder providers when increasing the laser power by 15% because of the better melting process of the particles.
- Quadratic roughness Sq presented similar behavior concerning the Sa, however, with slightly higher magnitudes (+29%) due to its more sensibility to topography heights' variation.
- Statistical and profile skewness Ssk and kurtosis Sku parameters indicate that the printed surface can be useful to cell viability (propagation, differentiation, and adhesion) given that skewness is positive kurtosis is greater than 3 (reference value), which means there is a prevalence of sharp peaks.

References

1. Yang, K.V., de Looze, G.R., Nguyen, V., Wilson, R.S.: Directed-energy deposition of Ti-6Al-4V alloy using fresh and recycled feedstock powders under a reactive atmosphere. Addit. Manuf. **58**, 103043 (2022)
2. Sivamani, S., Nadarajan, M., Kameshwaran, R., Bhatt, C.D., Premkumar, M.T., Hariram, V.: Analysis of cross-axis wind turbine blades designed and manufactured by FDM-based additive manufacturing. Mater. Today: Proc. **33**, 3504–3509 (2020)
3. Gibson, I., Rosen, D., Stucker, B., Khorasani, M.: Materials for Additive Manufacturing. In: Additive Manufacturing Technologies. Springer, Cham (2021). https://doi.org/10.1007/978-3-030-56127-7
4. Leary, M.: Surface roughness optimization for selective laser melting (SLM): Accommodating relevant and irrelevant surfaces. In: Brandt, M. (ed.) Laser Additive Manufacturing, pp. 99–118. Woodhead Publishing. (2017)
5. Jemat, A., Ghazali, M.J., Razali, M., Otsuka, Y.: Surface Modifications and Their Effects on Titanium Dental Implants. BioMed Res. Inter., 791725 (2015)
6. Oyelola, O., Crawforth, P., M'Saoubi, R., Clare, A.T.: On the machinability of directed energy deposited Ti6Al4V. Addit. Manuf. **19**, 39–50 (2018)
7. Azarniya, A., et al.: Additive manufacturing of Ti–6Al–4V parts through laser metal deposition (LMD): Process, microstructure, and mechanical properties. J. Alloys Compounds **804**, 163–191 (2019)
8. Kwak, T.Y., et al.: Additive manufacturing of a porous titanium layer structure Ti on a Co-Cr alloy for manufacturing cementless implants. J. Market. Res. **10**, 250–267 (2021)
9. Wong, M., Eulenberger, J., Schenk, R., Hunziker, E.: Effect of surface topology on the osseointegration of implant materials in trabecular bone. J. Biomed. Mater. Res. **29**(12), 1567–1575 (1995)
10. Selcuk, C.: Laser metal deposition for powder metallurgy parts. Powder Metall. **54**(2), 94–99 (2011)

11. Hacking, S.A., et al.: Surface roughness enhances the osseointegration of titanium headposts in non-human primates. J. Neurosci. Methods **211**(2), 237–244 (2012)
12. Ye, C., Zhang, C., Zhao, J., Dong, Y.: Effects of post-processing on the surface finish, porosity, residual stresses, and fatigue performance of additive manufactured metals: a review. J. Mater. Eng. Perform. **30**(9), 6407–6425 (2021)
13. Bagehorn, S., Wehr, J., Maier, H.J.: Application of mechanical surface finishing processes for roughness reduction and fatigue improvement of additively manufactured Ti-6Al-4V parts. Int. J. Fatigue **102**, 135–142 (2017)
14. Gockel, J., Sheridan, L., Koerper, B., Whip, B.: The influence of additive manufacturing processing parameters on surface roughness and fatigue life. Int. J. Fatigue **124**, 380–388 (2019)
15. Tolochko, N.K., et al.: Balling processes during selective laser treatment of powders. Rapid Prototyping J. **10**(2), 78–87 (2004)
16. Li, P., Warner, D.H., Fatemi, A., Phan, N.: Critical assessment of the fatigue performance of additively manufactured Ti–6Al–4V and perspective for future research. Int. J. Fatigue **85**, 130–143 (2016)
17. Greitemeier, D., Donne, C.D., Syassen, F., Eufinger, J., Melz, T.: Effect of surface roughness on fatigue performance of additively manufactured Ti–6Al–4V. Mater. Sci. Technol. **32**(7), 629–634 (2016)
18. Chan, K.S., Koike, M., Mason, R.L., Okabe, T.: Fatigue life of titanium alloys fabricated by additive layer manufacturing techniques for dental implants. Metall. and Mater. Trans. A. **44**(2), 1010–1022 (2013)
19. Svetlizky, D., et al.: Directed energy deposition (DED) additive manufacturing: Physical characteristics, defects, challenges and applications. Mater. Today **49**, 271–295 (2021)
20. Wennerberg, A., Albrektsson, T.: Suggested guidelines for the topographic evaluation of implant surfaces. Inter. J. Oral **15**(3), 331–344 (2000)
21. Zareidoost, A., Yousefpour, M., Ghaseme, B., Amanzadeh, A.: The relationship of surface roughness and cell response of chemical surface modification of titanium. J. Mater. Sci. - Mater. Med. **23**(6), 1479–1488 (2012)
22. Hatamleh, M.M., Wu, X., Alnazzawi, A., Watson, J., Watts, D.: Surface characteristics and biocompatibility of cranioplasty titanium implants following different surface treatments. Dent. Mater. **34**(4), 676–683 (2018)
23. Krishna Alla, R., Ginjupalli, K., Upadhya, N., Shammas, M., Krishna Ravi, R., Sekhar, R.: Surface roughness of implants: A review. Trends in Biomater. Artifi. Organs **25**(3), 112–118 (2011)
24. Carpenter Additive, Test Certificate, Chemical Analysis (2019)
25. AP&C (2020), Test Certificate, Chemical Analysis (2019)
26. Antonysamy, A.A.: Microstructure, texture, and mechanical property evolution during additive manufacturing of Ti6Al4V alloy for aerospace applications. The University of Manchester, United Kingdom (2012)
27. Averardi, A., Cola, C., Zeltmann, S.E., Gupta, N.: Effect of particle size distribution on the packing of powder beds: A critical discussion relevant to additive manufacturing. Mater. Today Commun. **24**, 100964 (2020)

Influence of Printing Parameters on Impact and Tension Resistance of Parts Produced by the Fused Filament Fabrication Process

Thiago de Oliveira Santos[1](✉), Kauã Ferreira de Almeida[1], Kenji Fabiano Ávila Okada[1], Felipe dos Anjos Rodrigues Campos[1], Leonardo Rosa Ribeiro da Silva[1], and Álisson Rocha Machado[2]

[1] Federal University of Uberlândia, 2121 João Naves de Ávila Avenue, Uberlândia, MG, Brazil
thiago.de@ufu.br

[2] Pontifical Catholic University, 1155 Imaculada Conceição Avenue, Curitiba, PR, Brazil

Abstract. Additive manufacturing by the Fused Filament Fabrication (FFF) process is a manufacturing process that has great potential for facilitating the production of complex and customized parts quickly, with low production costs and little material waste. However, the mechanical resistance of the parts produced by this manufacturing method vary greatly according to printing parameters, especially when they are not bulk solid. It was observed in this work how the impact and tensile resistance of a part change according to the thickness of the shell, height of the printed layer and the infill since these factors alter the moment of inertia and thermal loads. For instance, it is hard to predict if the mechanical properties will improve when increasing shell thickness and reducing infill, or vice versa. Specimens were then produced following a complete factorial design for these 3 variables at 2 levels, and then an analysis of variance was performed with the data from tensile strength tests according to ASTM D638 and impact resistance according to ASTM D4508. Shell thickness was shown to be the most influential and advantageous.

Keywords: Shell Thickness · Layer Height · Infill · Mechanical Strength · Moment of Inertia

1 First Section

In recent years, with the advancement of engineering, additive manufacturing has evolved so that it is no longer used only for rapid prototyping and 3D representation of designs, but also used as a manufacturing method capable of directly producing final parts. In this way, additive manufacturing has been benefiting several areas beyond mechanics such dentistry [1], medicine [2] and as architecture [3], with its capacity to produce parts in small quantity faster, simpler, cheaper and with less waste of material than other means of manufacturing, making it the best option for production of customized parts with a high level of complexity.

D. de Oliveira et al. (Eds.): COBEF 2023, LNME, pp. 161–171, 2024.
https://doi.org/10.1007/978-3-031-43555-3_15

Additive manufacturing by the Fused Filament Fabrication (FFF) process consists of extruding a fused filament through a mobile nozzle responsible for designing the part layer by layer using data from a 3D computational model as a basis. The materials most used in this method of production are thermoplastic polymers due to their melting point between 150 °C and 250 °C and satisfactory mechanical properties at room temperature, with some examples of widely used materials being ABS, PLA and PETG.

Due to the part being printed layer by layer, its mechanical properties are directly related to the adhesion between the deposited filaments and between the layers formed by them, and the lack of adhesion might facilitate the nucleation and propagation of cracks that cause material premature failure [4]. Thus, parts produced by additive manufacturing have strong anisotropy, with reduced resistance in the build direction that led to failures in tensile tests [5, 6], torsion [7], compression [8, 9] and flexural tests [10]. For this reason, this remains one of the major obstacles and research topics in additive manufacturing.

The adhesion between the extruded polymeric filaments is based on the molecular diffusion of the polymer chains from one filament to the other and the intertwining between them, and this diffusion depends on the thermal history in the printed part [10]. When the filament is extruded, it loses thermal energy to the environment and to the other layers, and the longer it remains at a temperature above the glass transition temperature of the polymer, more the polymer chains will diffuse between the extruded filaments and intertwine. In addition, a higher temperature induces greater fluidity and wettability of the deposited material, which consequently generates a larger contact area between the filaments, as shown in Fig. 1a. This effect is visible in Fig. 1b, which also shows how the spaces between the deposited filaments are smaller in the bottom layers, which receive heat from the printing table for longer, ensuring more time for diffusion and adhesion of the layers.

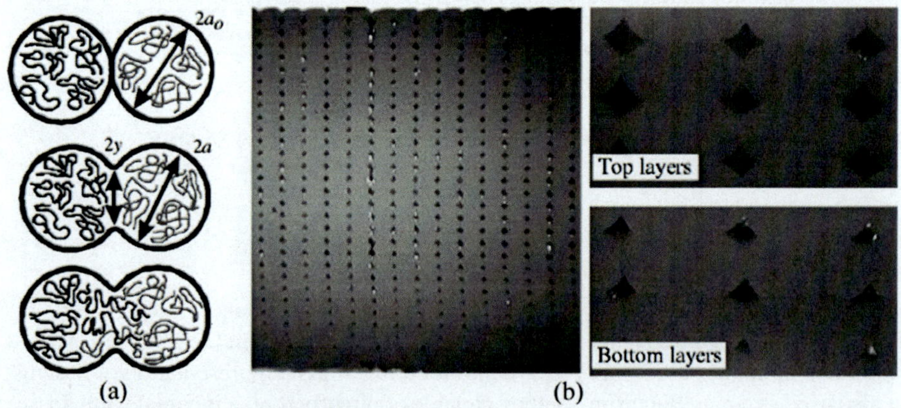

Fig. 1. (a) Mechanism of adhesion between the deposited layers, dependent on the intertwining of the polymeric chains. (b) Larger voids and lesser contact area between filaments in the top layers compared to the bottom layers. Adapted [10].

Due to this phenomenon, many researchers aim to increase the mechanical strength by other means, leaving the part less vulnerable to the thermal effects intrinsic of this manufacturing method. When designing a part to be printed, attention should be paid to the infill and the thickness of the part's shell, as one of the benefits of 3D printing is precisely the ability to produce parts without necessarily being massive. These two parameters directly influence the moment of inertia of the designed part, which is given by the integral sum of the mass of each point of the part times the square of the distance to the centroid. Thus, considering the rectangular cross section of a tensile test specimen for instance, theoretically the regions furthest from the center contribute greater for the resistance. Therefore, a greater wall thickness and increased bottom and top layer numbers may be more advantageous than a larger infill to increase mechanical strength.

In addition to these two parameters, another important parameter to evaluate when designing a part is the layer height, as observed in the literature [11–13]. It is known that a lower layer height is related to an increase in the mechanical resistance of the manufactured parts, in addition to resulting in a better finish and decreasing the interstitial spaces between the filaments shown in Fig. 1b, which consequently increases the density and mechanical resistance. However, it must also be considered that a lower layer height, just like a higher percentage of infill and thicker shell increases considerably the production time, while the last two parameters also increase the filament consumption.

In this way, as each of these parameters has its advantages and disadvantages, it necessary to study how influential they are to allow optimization of the mechanical properties, printing time and filament consumption of parts produced by the FFF process. This research employed a complete factorial design, with three variables at two levels each, to produce the tensile test and impact test specimens. It was found that the combination of these factors may lead to unforeseen results, such as lesser mechanical resistance when increasing infill. It was explained how the printing time could lead to this behavior, once it affects the thermal history of the samples.

2 Methodology

2.1 Sample Preparation

The impact and tensile tests were carried out with samples of polylactic acid (PLA), and the filament used was uncolored PLA from 3DFila. This is important because other works have shown that pigment loading may affect mechanical resistance of the polymer [14]. The specimens were produced FFF using a Creality Ender 3 3D printer and Ultimaker Cura Slicer Software. The input parameters evaluated were layer height, shell thickness and infill. Shell thickness is not a readily defined parameter in Cura, which allows for changing separately the wall thickness, bottom layer number and top layer number. The values were chosen so that a regular thickness could be obtained in all sides of the samples. The output parameters were the energy absorbed by the sample, the maximum tensile strength, the printing time, and filament consumption, with the last two as yielded by the printer in the slicing software. The full factorial design resulting in the 8 printing conditions is presented in Table 1.

The parameters were chosen with a difference of 50% (1.2 mm/0.8 mm of shell thickness and 60%/40% of infill) as it is already known that a much greater infill or shell thickness would increase the mechanical strength. Therefore, the intention was to observe whether small changes in these parameters would affect the results, and the

Table 1. Factorial experimental design 2^3.

Condition	Layer height [mm]	Shell thickness [mm]	Infill [%]
1	0.1	0.8	40
2	0.1	0.8	60
3	0.1	1.2	40
4	0.1	1.2	60
5	0.2	0.8	40
6	0.2	0.8	60
7	0.2	1.2	40
8	0.2	1.2	60

significance of one parameter in relation to the others. It is noteworthy that the layer height did not change by 50%. In this case, since the test specimen thickness of 3.2 mm is suggested by the standards, this number should be a multiple of both chosen layer height values. Otherwise, the 3D slicing software could round off the values and affect the final dimension of the specimen.

For each printing condition presented in Table 1, three specimens were produced for for the two mechanical tests performed, totaling 48 specimens produced. Although ASTM D256 and ASTM D638 recommend evaluating at least 5 samples for each test condition, there were time and resource constraints during sample fabrication.

Although printing several samples together would be faster, each part was printed individually to standardize the heat received from the table for each sample after it is printed. This is especially important because the higher thermal intake could provide more time for the polymer chains to diffuse and lead to inaccuracies in the results. In addition, the other printing parameters were fixed for all specimens, being equal to the default settings of the slicer software configured for this printer used and for generic PLA. For instance, the temperature of the extruder nozzle and the temperature of the printing table were, respectively, 200 °C and 50 °C.

2.2 Impact Tests

The Izod type C impact tests were carried out in accordance with the recommendations of the ASTM D256 standard, using specimens with the dimensions shown in Fig. 2. It is also worth noting that the impact specimens were printed with the orientation of the layers as shown in Fig. 3. Although the standard states that the notch should be machined, it was already defined in the CAD model for printing. Since it might affect the results due to differences in residual tensions and surface integrity [15], the results should only be compared in between the conditions, and not to samples with machined notches from other works.

The impact tests were performed on an impact testing machine adapted from other work [16], using a pendulum with a mass of 0.99 kg and a length of 0.778 m, which resulted in a pendulum energy of 1.79 J, releasing it from rest from an inclination

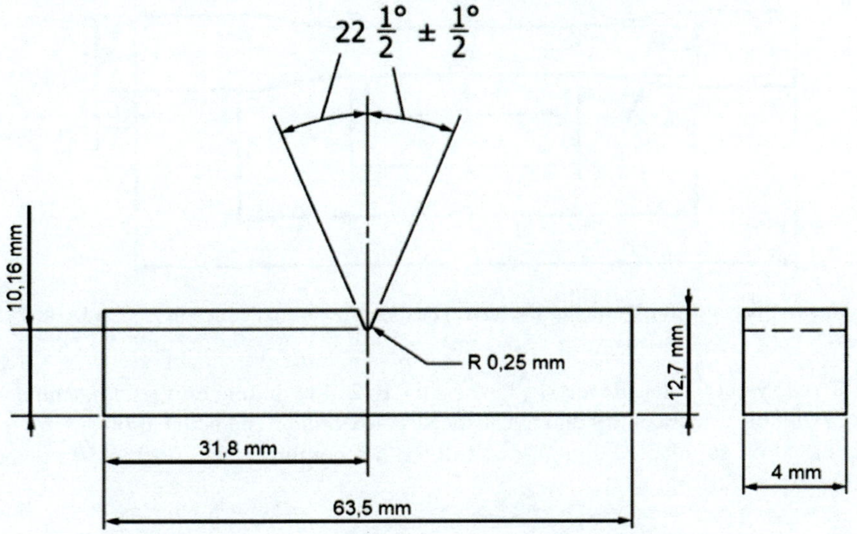

Fig. 2. Example of specimen for Izod type C tests (ASTM D256).

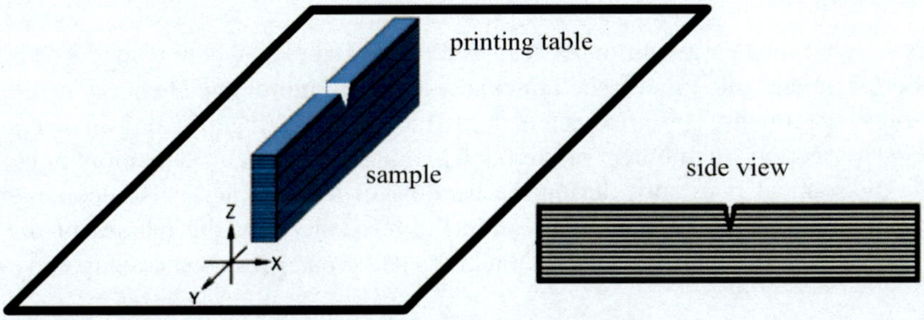

Fig. 3. Orientation of layers used in impact specimens. Adapted from Patterson et al. (2019).

of −30° with the horizontal, for all tests. Under these conditions, the tip of the bar impacted the samples with a velocity of approximately 4 m/s, which provides an idea of the shear rate on the material.

2.3 Tensile tests

The tensile tests were carried out following the recommendations of ASTM D638-14 standard, with 4 mm thick V-type specimens, as shown in Fig. 4. However, unlike the impact specimens, the tensile specimens were printed with their largest side parallel to the printing table. Even so, it is noteworthy that in both tests the main forces act in the longitudinal direction of the sample, aligned with the direction of the deposited layers. This is evident in the tensile test, and can also be seen in the impact test, where the fracture results from the flexion and stretching effect of the layers.

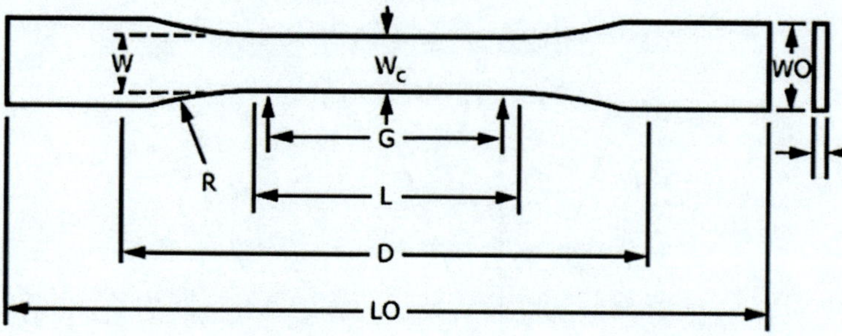

Fig. 4. Specimen for V-type tensile tests. The dimensions are defined in ASTM D638-14.

To carry out the tensile tests, a Pavitest 5T I-4214 hydraulic universal machine was used, with electrical control and a test speed of 0.1 mm/s. The acquisition system used was based on an adapted digital load cell, with a nominal range of 0 to 50 kg and a resolution of 0.01 kg.

3 Results and Discussion

3.1 Impact Tests

Figure 5 shows the energy of impact absorbed by the samples, where it can be observed that the specimens with greater shell thickness absorbed more impact energy in relation to the other specimens. This is expected, due to the thicker shell generating an increase in the resistant section and moment of inertia, with the outer regions being more important for the mechanical resistance during the bending of the sample. This becomes more evident when observing that even dividing the resistances by the masses of the test specimens, the parts with greater shell thickness had greater resistance compared to the others.

Contrary to expected, the infill did not show major changes, which may be related to how the load is distributed when applied to the specimen. It is likely that the impact of the pendulum generated greater tensions in the outer region of the part, with the filling being less relevant for part resistance.

The influence of the layer height was different from what was expected, although it was not so significant. Instead of increasing the mechanical resistance with a lower layer height, a decrease in the resistance of some parts is observed. Initially, it was expected that a lower roughness together with smaller interstitial spaces, resulting from a lower layer height, would yield less stress concentration and better adhesion between layers due to prolonged exposure to heat, considering the longer production time of these parts. However, the higher thermal intake might have been enough to make the part less ductile due to the crystallization of the polymer chains [17], which is negative for impact strength, where ductility is important. It can be observed that the production time of the parts with lower layer height is almost twice as long as the others, which means that the material spends much more time receiving heat from the nozzle and the printing table, which could be enough to change the properties. of a low glass transition temperature thermoplastic such as PLA [4].

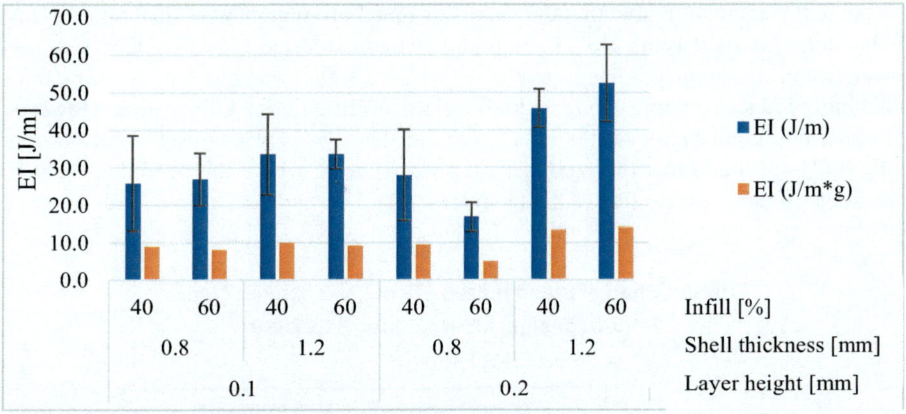

Fig. 5. Variation of energy of impact (EI) absorbed by the samples for each printing condition.

Table 2. Printing time for each combination of parameters in the impact test.

Layer height [mm]	Shell thickness [mm]	Infill [%]	Print time [min]
0.1	0.8	40	52
0.1	0.8	60	59
0.1	1.2	40	59
0.1	1.2	60	66
0.2	0.8	40	26
0.2	0.8	60	30
0.2	1.2	40	30
0.2	1.2	60	33

In addition, the fact that a higher layer height did not worsen the impact resistance may be related to the build print being perpendicular to the impact direction, which would make the roughness effect insignificant, since the notch already served as a much more relevant stress concentration point. In the graph it can also be seen that in some cases with similar parameters, where the infill was the only difference, parts with 40% infill had better results, which differs from the results found in the literature and can also be explained by this greater temperature exposure time, due to longer production time with longer infill. Table 2 presents the printing time for each combination of parameters of the impact specimens.

These observations are in line with the effects observed in Fig. 6, obtained by analysis of variance (ANOVA), performed with Statistica 12 software. The data shows that shell thickness and the combination of shell thickness with layer height are the only factors that were significant for a confidence interval of 95%, and shell thickness has a much greater significance than the other factors and combinations.

Also, it is interest to note that the results of Fig. 5 are much lower than other results in the literature, such as the 113.21 J/m found by Patterson et al. (2019) in Izod tests performed on PLA. Again, this can be explained by the difference in notch fabrication, since machining induces compression residual stress that difficult crack formation. However, the most important factor in this case would be that these authors used solid samples with 100% infill and that the Izod test used was type E, where the pendulum hits the specimen on the opposite side to the chamfer.

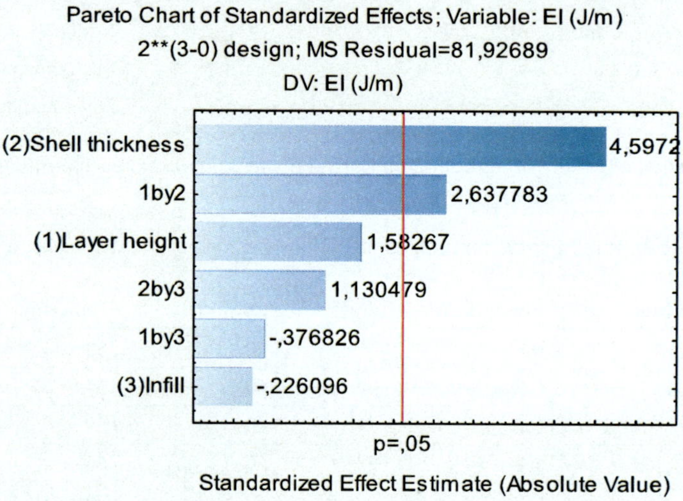

Fig. 6. Pareto chart from ANOVA showing significance of factors for energy of impact (EI).

3.2 Tensile Tests

The variations in the maximum tensile strength for the different printing conditions are shown in Fig. 7 and are within the range of values expected for PLA in the literature [18]. It is evident that all test specimens with greater shell thickness presented higher rupture stress, as previously observed with the impact tests. This is expected, since the tensile test was carried parallel to main axis of the part, in which shell filaments are aligned. This increases the resistance due to a higher moment of inertia, regardless of the adhesion between layers. This observation is reinforced with the results obtained from ANOVA, which are shown in Fig. 8, confirming that the shell thickness really has a much greater influence compared to the other parameters.

It can also be seen that shell thickness and its combination with the layer height were again the only statistically significant factors for a confidence interval of 95%. From the graph in Fig. 7, it can also be noted that many parts with 40% infill showed better results compared to conditions with 60% infill and the same layer height and shell thickness, which is counter intuitive at first. In Table 3 it can be observed that the printing time

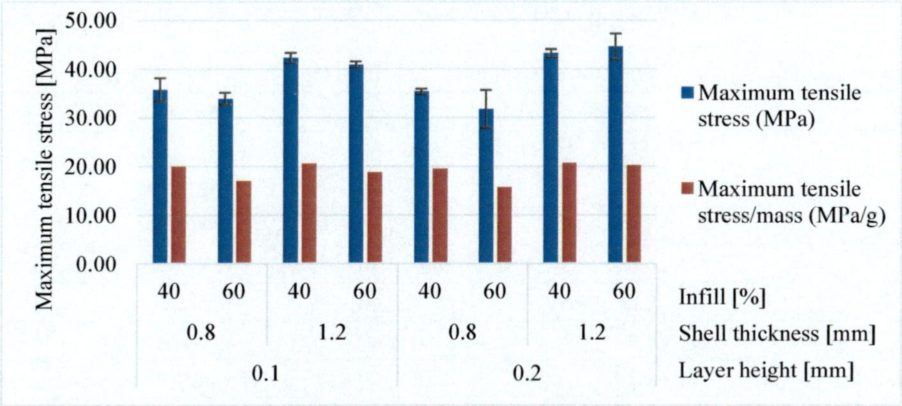

Fig. 7. Variation of maximum tensile stress for each combination of variables.

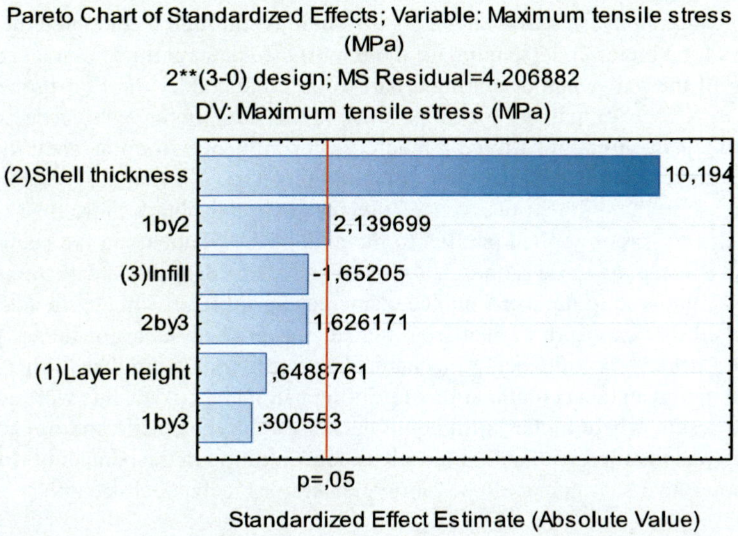

Fig. 8. Pareto chart showing significance of factors for the maximum tensile stress.

was slightly shorter for the more resistant parts, which may indicate that exposure to temperature may have again impaired the results in the tensile tests, as well as in the impact tests. It is also possible that, since the thickness of the shell is more important for the strength of the part, the infill could have a negative effect, with the contact points between the filling and the shell serving as stress concentration points that facilitate crack nucleation and propagation.

Table 3. Printing time for each combination of parameters in the tensile test.

Layer height [mm]	Shell thickness [mm]	Infill [%]	Print time [min]
0.1	0.8	40	28
0.1	0.8	60	31
0.1	1.2	40	33
0.1	1.2	60	35
0.2	0.8	40	14
0.2	0.8	60	15
0.2	1.2	40	17
0.2	1.2	60	18

4 Conclusion

The significance analysis, based on the results obtained through the adopted methodology, allows for a better understanding of the importance of shell thickness, layer height and filling of the part when designing a part to be produced by the FFF process, and may also serve as a comparison for future work in this area. In the tests carried out, the impact and tensile strength suffered a much greater influence from the shell thickness than for the other parameters and their combinations. One of the main factors for this is the increase in the moment of inertia, since the shell contributes more than the bulk for resistance to forces applied parallel to the axis of the filaments in the shell layers. This shows a faster and more economical way of increasing 3D printed parts mechanical properties, compared to the usual choice of increasing infill percentage. In addition, it was observed that exposure to heat from nozzle and printing table for longer periods may end up negatively influencing mechanical strength, indicating that further studies are needed to find an ideal combination of printing parameters. Also, this work found an interesting result, where higher infill could decrease both the tensile and impact resistance. This was justified with a possible stress concentration in the contact of the filling with the outer layers, which could be more relevant for mechanical strength.

References

1. Bhargav, A., Sanjairaj, V., Rosa, V., Feng, L.W., Fuh, Y.H.J.: Applications of additive manufacturing in dentistry: a review. J. Biomed. Mater. Res. Part B Appl. Biomater. **106**(5), 2058–2064 (2018)
2. Konta, A.A., García-Piña, M., Serrano, D.R.: Personalised 3D printed medicines: which techniques and polymers are more successful? Bioengineering **4**(79), 1–16 (2017)
3. Haridy, R.: Get over it: Madrid gets world's first 3D printed footbridge. https://newatlas.com/3d-printed-bridge-madrid/47650/. Accessed 23 May 2023
4. Sharafi, S., Santare, M.H., Gerdes, J., Advani, S.G.: A review of factors that influence the fracture toughness of extrusion-based additively manufactured polymer and polymer composites. Addit. Manuf. **38**(101830), 1–24 (2021)

5. Cole, D.P., Riddick, J.C., Iftekhar Jaim, H.M., Strawhecker, K.E., Zander, N.E.: Interfacial mechanical behavior of 3D printed ABS. J. Appl. Polym. Sci. **133**(30), 1–12 (2016)
6. Fernandez-Vicente, M., Calle, W., Ferrandiz, S., Conejero, A.: Effect of infill parameters on tensile mechanical behavior in desktop 3D printing. 3D Print. Addit. Manuf. **3**(3), 183–192 (2016)
7. Rodríguez, J.F., Thomas, J.P., Renaud, J.E.: Mechanical behavior of acrylonitrile butadiene styrene (ABS) fused deposition materials. Experimental investigation. Rapid Prototyp. J. **7**(3), 148–158 (2001)
8. Percoco, G., Lavecchia, F., Galantucci, L.M.: Compressive properties of FDM rapid prototypes treated with a low cost chemical finishing. Res. J. Appl. Sci. Eng. Technol. **4**(19), 3838–3842 (2012)
9. Sood, A.K., Ohdar, R.K., Mahapatra, S.S.: Experimental investigation and empirical modelling of FDM process for compressive strength improvement. J. Adv. Res. **3**(1), 81–90 (2012)
10. Sun, Q., Rizvi, G.M., Bellehumeur, C.T., Gu, P.: Effect of processing conditions on the bonding quality of FDM polymer filaments. Rapid Prototyping J. **14**(2), 72–80 (2008)
11. Percoco, G., Arleo, L., Stano, G., Bottiglione, F.: Analytical model to predict the extrusion force as a function of the layer height, in extrusion based 3D printing. Addit. Manuf. **38**(101791), 1–9 (2021)
12. Dey, A., Yodo, N.: A systematic survey of FDM process parameter optimization and their influence on part characteristics. J. Manuf. Mater. Process. **3**(64), 1–30 (2019)
13. Mahmood, S., Qureshi, A.J., Goh, K.L., Talamona, D.: Tensile strength of partially filled FFF printed parts: experimental results. Rapid Prototyping J. **23**(1), 122–128 (2017)
14. Wittbrodt, B., Pearce, J.M.: The effects of PLA color on material properties of 3-D printed components. Addit. Manuf. **8**(1), 110–116 (2015)
15. Patterson, A.E., Pereira, T.R., Allison, J.T., Messimer, S.L.: IZOD impact properties of full-density FDM polymer materials with respect to raster angle and print orientation. Proc. IMECHE Part C J. Mech. Eng. Sci. **235**(10), 1891–1908 (2019)
16. Tsuruta, K.M.: Monitoramento de integridadeestrutural de materiais compostos sujeitos a impactos empregando a técnica da impedância eletromecânica. Master's dissertation, Federal University of Uberlândia, Uberlândia (2008)
17. Samy, A.A., Golbang, A., Harkin-Jones, E., Archer, E., Tormey, D., McIlhagger, A.: Finite element analysis of residual stress and warpage in a 3D printed semi-crystalline polymer: effect of ambient temperature and nozzle speed. J. Manuf. Process. **70**(1), 389–399 (2021)
18. Hanon, M.M., Marczis, R., Zsidai, L.: Influence of the 3D printing process settings on tensile strength of PLA and HT-PLA. Periodica Polytechnica Mech. Eng. **65**(1), 38–46 (2021)

Heat Treatment Influence on Micromilling of Additively Manufactured Titanium

Maria Clara Coimbra Gonçalves[1,3](✉), Marcello Vertamatti Mergulhão[2], Gilmar Ferreira Batalha[1], and Rodrigo Lima Stoeterau[1]

[1] USP - Escola Politécnica da Universidade de São Paulo, Cidade Universitária, Campus Butantã, São Paulo, SP, Brazil
mclara.coimbra@usp.br

[2] IPEN - Instituto de Pesquisas Energéticas e Nucleares, Cidade Universitária, Campus Butantã, São Paulo, SP, Brazil

[3] Mechanical Engineering Department, University of Sheffield, Sheffield, UK

Abstract. Recently, miniaturized products are being used as working tools by many fields, highlighting the medical and dental industries. With the development of Additive Manufacturing (AM) processes, these features can be produced with less material waste and with wide design possibilities by applying a 'near-net-shape' technique. Although, the AM parts require post-processing techniques to reach the required material properties, dimensions, and surface roughness and to reduce undesirable residual stresses. In this aspect, the micromilling process is usually applied to attain the desired dimensions and surface roughness and heat treatments to attain the desired material properties and to reduce residual stress. Nonetheless, the micromilling process can be done before or after the heat treatment. In this perspective, this research analyses the surface roughness results for the micromilled AM parts before and after the heat treatment, which is in important for planning the manufacturing route for these parts. Thus, this work aims to compare the surface roughness results of *Sa*, *Sq*, *Ssk*, *Sku* when performing the micromilling process on AM parts by Laser Powder Bed Fusion (LPBF), with and without heat treatment. For the experiments, the tool size, feed and cutting speed were varied in a full factorial design of experiments. After that, the surface roughness parameters were analyzed and compared for both workpieces. With the achieved results, it can be concluded that the surface texture (*Ssk*) for the heat-treated and non-heat-treated samples present a predominance of peaks. Also, there is a presence of inordinately high peaks and/or deep valleys on the surface (*Sku*), which can present an interference of the chips left on the surface not removed by the ultrasonic cleaning. By the analyses of the arithmetic mean deviation (*Sa*) and the root mean square height (*Sq*), a better surface quality was achieved when micromilling the samples before the heat treatment for greater tool size, for the specific set of parameters used. With the smaller tool size, a greater surface roughness was achieved if compared to the bigger tool size, though the difference between the samples were not expressive. Moreover, the results achieved in this work can be applied to improve the surface quality of the AM parts used in industry.

Keywords: Machining · Micromilling · Additive manufacturing · Titanium · Heat treatment

D. de Oliveira et al. (Eds.): COBEF 2023, LNME, pp. 172–182, 2024.
https://doi.org/10.1007/978-3-031-43555-3_16

1 Introduction

Titanium and its alloys are commonly applied for manufacturing microcomponents for medical, odontological, aerospace and aeronautical industries due to their high corrosion resistance, excellent mechanical strength, and low density [1]. Apart from the conventional machining, additive manufacturing (AM) is an alternative process that can be applied in order to optimize the production of these microparts. Among its benefits, Uhlmann *et al.* 2015 [2] and Chen *et al.* 2017 [3] highlight its lower resource consumption, complex shapes production, less material waste, and the possibility of a rapid manufacturing in a variety of materials.

Even with these benefits, applying AM strategies on metallic materials lead to high surface roughness and residual stresses on the outcomes, what can be improved with micromilling process and heat treatments, respectively. In this perspective, Campos *et al.* 2020 [4] made a comparison on the micromilling machinability of commercial and AM Ti6Al4V workpieces. The authors considered cutting forces, surface roughness, burr formation and chip morphology for the analysis, though the residual stresses on the surface or a heat treatment effect was not analyzed. In the same area, Cardoso and Davim 2010 [5] aimed to optimize the surface roughness on the microsurface of an aluminum alloy by observing the relationship of *Ra* and feed rate. For the optimization, the authors used an analytical relationship between two roughness average and machining time. Although, in their work, the sample was conventionally manufactured, and the surface roughness of an area was not analyzed.

Regarding the microstructure effects of heat treated samples on the micromilling process, Aksin and Karpat 2019 [6] investigated the machining forces and surface roughness while micromachining commercially pure titanium. The authors observed that as the sample's microstructure becomes more equiaxed, its hardness increases, what also increases the machining forces. Although, the surface roughness gets better with increasing forces, as a possible decrease in the material ductility. On the other hand, Ahmadi *et al.* 2018 [7] analyzed the effect of heat treatment on a commercial Ti6Al4V alloy and found that the surface roughness was better for the microstructure with lower hardness and fine equiaxed grains. This comparison confirms the statement that there is a complex interplay between process parameters and process outputs during micromilling [6].

In summary, different material microstructures can directly affect the micromilling process quality [8]. For this reason, this paper aims to contribute with literature, as no previous work was found comparing the effect of heat treatment on surface roughness results when micromilling LPBF manufactured Ti6Al4V alloy. In this perspective, the present work aims to analyze some of the surface texture field parameters (*Sa*, *Sq*, *Sku* and *Ssk*), since these parameters are being increasingly applied in industries as a sophisticated method to address surface functionality of parts [9].

2 Materials and Methods

2.1 Additive Manufacturing Process

For manufacturing the workpieces used in this work, an OmniSint-160 LPBF equipment from the Brazilian company Omnitek was used. For the AM process, an Ytterbium fiber laser with a maximum power of 500 W and a 140 μm diameter laser beam focus was applied under Argon atmosphere. The process parameters used are summarized in Table 1. The measured density of the Ti6Al4V powder was 4.39 ± 0.001 g/cm^3, which was obtained by Helium pycnometry at the Nuclear Fuel Center – CCN/IPEN using the Micromeritics equipment – AccuPyc 1330 Pycnometer model. To analyze the powder granulometry, a particle analyzer equipment (Cilas Particle Size Analyzer - Model 1064 - from CCTM IPEN) using the Fraunhofer method was applied for measuring the particles average size. The results led to an average particle diameter of 43.47 μm.

After the AM process, some samples were submitted to an annealing heat treatment at 950 °C for 60 min. As studied by Etesami *et al.* 2022 [10], the annealing of an AM Ti6Al4V sample at lower transus temperature (900 °C) results in a decomposition of α' to α and β phases. In addition, it was shown that in this case the ultimate tensile strength is higher for the samples without heat treatment [10].

Table 1. Values of parameters used for the experiments in each material.

Laser power	Scanning speed	Layer thickness	Laser spot diameter	Scan spacing
155 W	950 mm/s	30 μm	140 μm	70 μm

2.2 Micromilling Process

The micromilling tests consisted in manufacturing slots on the workpieces with TiAlN coated tungsten carbide (WC) endmills with different set of parameters. The parameters used are shown in Table 2, and design of experiments applied was a full factorial design. The experiments were performed in a CNC Mini-mill/GX by Minitech Machinery Corporation with a motor model EM-3060 from Nakanishi brand. This machine presents a positioning error of 0.1 μm, a maximum feed rate of 1000 mm/min and a maximum spindle speed of 60000 rpm. Before the experiments the surface was flattened with a 2 mm diameter mill. After the performing the main experiments, one replica was made. An illustration of the experimental setup is shown in Fig. 1.

Table 2. Cutting parameters of the micromilling process.

Cutting speed	60 m/min
Feed	1, 1.5, 2 μm/tooth
Tool diameter	0.5, 0.8 mm

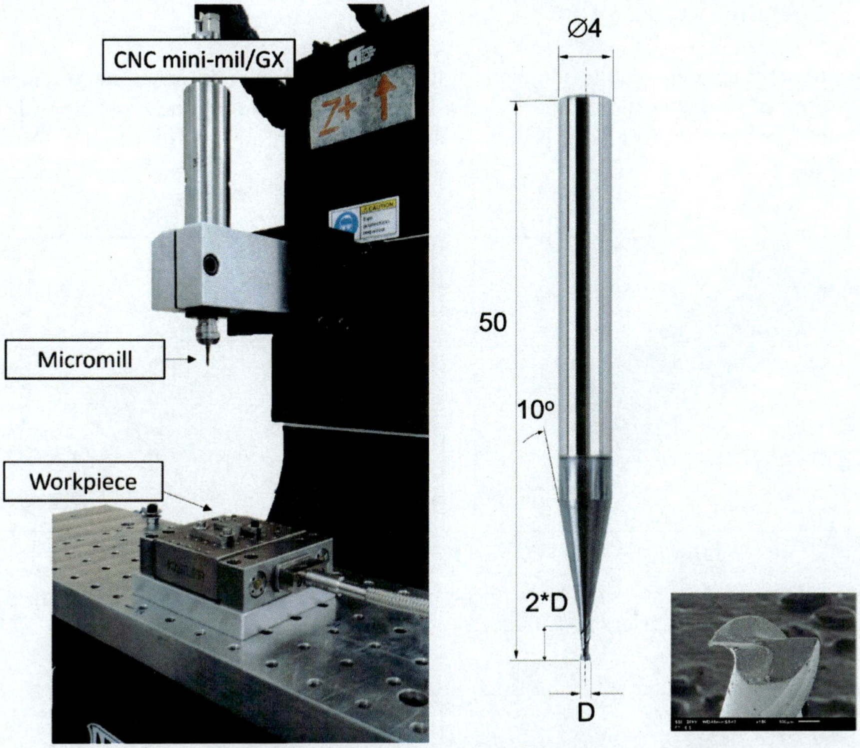

(a) CNC Mini-mill/GX (b) WC carbide micromill

Fig. 1. Experimental setup.

After the experiments the surface roughness parameters for each slot experiment were measured three times in a 3D CCI optical interferometer from Taylor Hobson. For the measurements, the parameters of *Sa* and *Sq* were considered, which represents the arithmetic mean deviation and the root mean square height of an area, respectively. The skewness Ssk and kurtosis Sku parameters were also measured, as they are related to the height distribution and to the peaks and valleys tip geometry, respectively [11]. Some conclusions can be made from the considered surface roughness parameters [11–13]:

- *Sa* cannot represent the punctual roughness profile, as it is an average value;
- *Sq* is more sensitive than *Sa* to large height variations;
- When the *Ssk* parameter is positive, there is a predominance of peak on the surface. While negative, there is a predominance of valleys;
- The *Sku* values above 3 represents a large concentration of amplitudes on the surface. While values below 3 indicate the absence of extreme peaks and valleys.

3 Experimental Results

The surface roughness results obtained for the heat treated and non-heat treated samples are presented in this section. Figure 2 and Fig. 3 **show a 3D roughness profile of both** samples after the micromilling process using the 0.5 mm and 0.8 mm tool diameter, respectively.

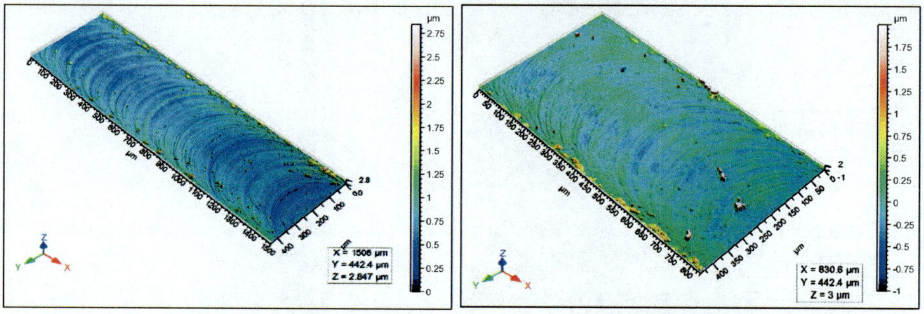

(a) Sample without HT (b) Sample with HT

Fig. 2. 3D surface roughness results for both samples under cutting speed of 60 m/min, feed of 1 μm/tooth and tool diameter of 0.5 mm.

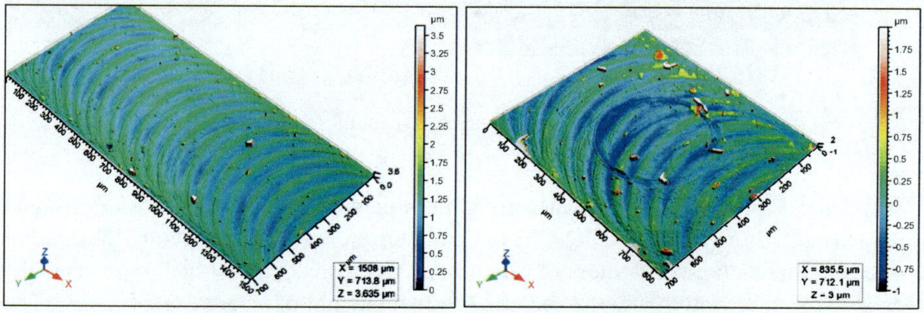

(a) Sample without HT (b) Sample with HT

Fig. 3. 3D surface roughness results for both samples under cutting speed of 60 m/min, feed of 1 μm/tooth and tool diameter of 0.8 mm.

The *Sa*, *Sq*, *Sku* and *Ssk* results obtained for the experiments are plotted in Fig. 5, Fig. 6, Fig. 7, and Fig. 8, respectively. From these pictures, it is possible to observe that due to size effect the results had a different behavior when tool size was changed. The *Sa* and *Sq* values, for example, were greater for the sample with heat treatment for the higher tool diameter. This behavior was not expected, since there is a reduction on the sample's hardness and ultimate tensile strength after the heat treatment. Although, since the heat treatment reduces the brittleness of the material, it favors built-up edge and continuous chip formation. Indeed, built-up edge and continuous chip formation were observed during these experiments (Fig. 4). Another observation of continuous chip formation worsening the surface roughness can be seen in Fig. 3b. In this case, the blue ring on the surface was possibly formed due to chip being dragged on the surface. These observations can explain the higher *Sa* and *Sq* values obtained for the higher tool diameter.

To better determine the effect of the tool size, feed per tooth and heat treatment on the surface roughness results, an analysis of variance (ANOVA) was performed for each result separately. Table 3, Table 4, Table 5, and Table 6 shows these results. As it can be observed from the ANOVA results, tool size and heat treatment parameters and its interaction had a statistically significant influence on Fig. 5, Fig. 6, Fig. 7, and Fig. 8 results. This is concluded from the P-value lower than 0.05 from these sources. The higher F-values indicate that the influence is not likely due by chance. In this way, feed per tooth influence and its interactions are inconclusive due to their P-value greater than 0.05 and lower F-value for all surface roughness parameters analysed.

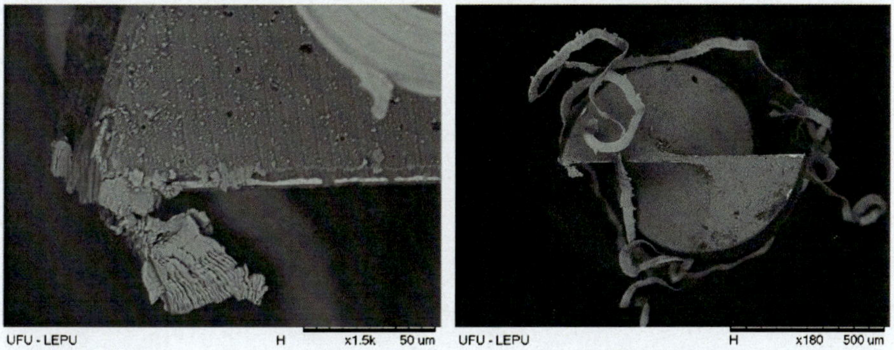

(a) Material adhered in the tool tip (b) Continuous chip tangles in the tool

Fig. 4. Images of the 0.8 mm tool diameter.

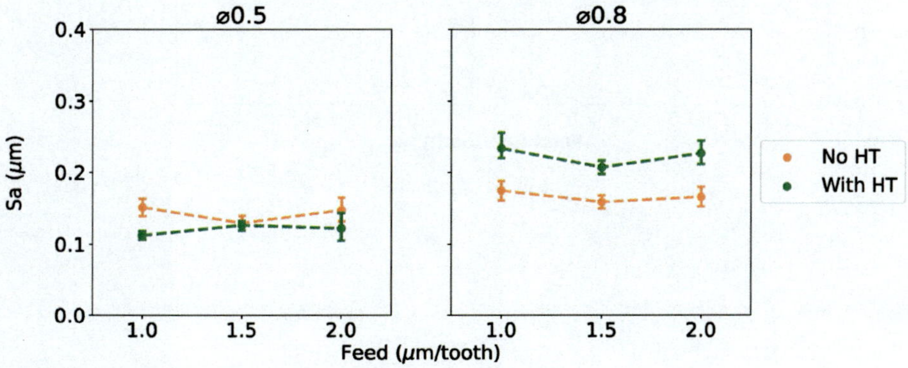

Fig. 5. *Sa* results

Observing the surface's morphology results (Fig. 7 and Fig. 8), it can be concluded that there is a peak predominance on the surface, as *Ssk* results were greater than zero for all experiments. Additionally, the *Sku* results greater than three for the experiments show that the surface presents extreme peaks and valleys, what can be related to the chips left on the surface even after ultrasonic cleaning.

Table 3. ANOVA results for analyzing *Sa*.

Source	DoF	Adj SS	Adj MS	F	P
Tool	1	0.030953	0.030953	134.96	0.000
Feed	2	0.001593	0.000796	3.47	0.065
Heat treatment	1	0.002260	0.002260	9.85	0.009
Tool*Feed	2	0.002531	0.001266	5.52	0.020
Tool*Heat treatment	1	0.010732	0.010732	46.79	0,000
Feed*Heat treatment	2	0.000620	0.000310	1.35	0.296
Tool*Feed*Heat treatment	2	0.000391	0.000196	0.85	0.450
Error	12	0.002752	0.000229		
Total	23	0.051832			

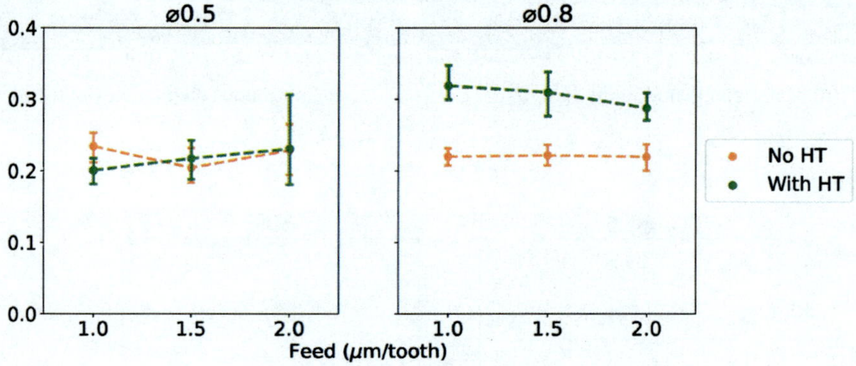

Fig. 6. *Sq* results

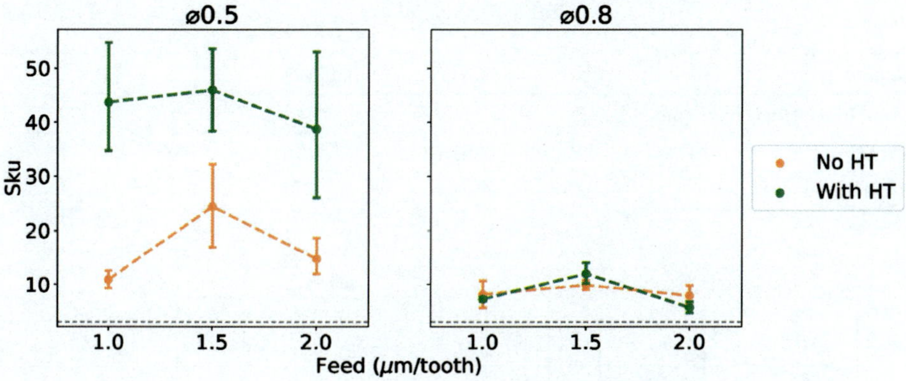

Fig. 7. *Sku* results

Table 4. ANOVA results for analyzing *Sq*.

Source	DoF	Adj SS	Adj MS	F	P
Tool	1	0.016627	0.016627	22.32	0.000
Feed	2	0.004006	0.002003	2.69	0.108
Heat treatment	1	0.009976	0.009976	13.39	0.003
Tool*Feed	2	0.003107	0.001554	2.09	0.167
Tool*Heat treatment	1	0.017501	0.017501	23.50	0.000
Feed*Heat treatment	2	0.001213	0.000607	0.81	0.466
Tool*Feed*Heat treatment	2	0.001567	0.000783	1.05	0.379
Error	12	0.008937	0.000745		
Total	23	0.062934			

Table 5. ANOVA results for analyzing *Sku*.

Source	DoF	Adj SS	Adj MS	F	P
Tool	1	3414.0	3413.98	46.79	0.000
Feed	2	146.4	73.21	1.00	0.395
Heat treatment	1	1635.5	1635.55	22.42	0.000
Tool*Feed	2	7.43	3.71	0.05	0.951
Tool*Heat treatment	1	1639.41	1639.41	22.47	0.000
Feed*Heat treatment	2	130.44	65.22	0.89	0.435

(*continued*)

Table 5. (*continued*)

Source	DoF	Adj SS	Adj MS	F	P
Tool*Feed*Heat treatment	2	114.86	57.43	0.79	0.477
Error	12	875.51	72.96		
Total	23	7963.6			

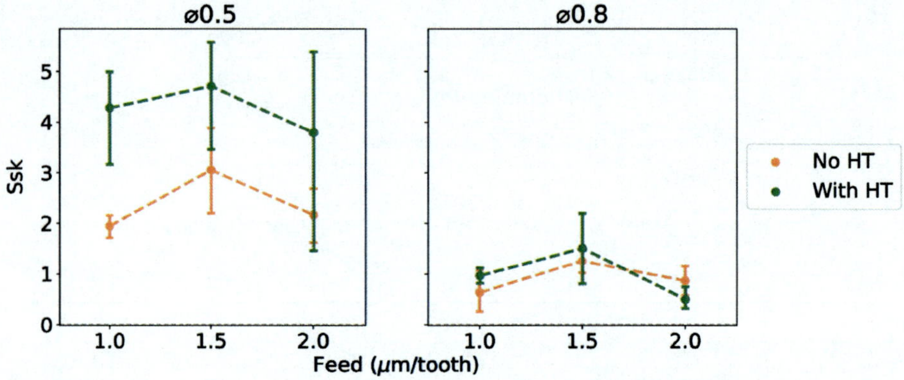

Fig. 8. *Ssk* results

Table 6. ANOVA results for analyzing *Ssk*.

Source	DoF	Adj SS	Adj MS	F	P
Tool	1	44.155	44.1545	72.90	0.000
Feed	2	3.562	1.7809	2.94	0.091
Heat treatment	1	12.884	12.884	21.27	0.001
Tool*Feed	2	0.4020	0.2010	0.33	0.724
Tool*Heat treatment	1	13.4291	13.4291	22.17	0.001
Feed*Heat treatment	2	1.6190	0.8095	1.34	0.299
Tool*Feed*Heat treatment	2	0.5472	0.2736	0.45	0.647
Error	12	7.2680	0.6057		
Total	23	83.866			

4 Conclusions

This article addressed a comparison of the surface roughness results regarding the heat treatment of an additively manufactured Ti6Al4V alloy. From the results obtained, the following conclusions can be drawn:

- There is an influence of the heat treatment on surface roughness results, though no link could be established between the different tool sizes results due to size effect;
- Tool size and heat treatment condition are important parameters in micromilling and can be used in process optimization;
- For the tool of 0.8 mm diameter the *Sa* and *Sq* results were higher for the sample with heat treatment, what was possibly influenced by the ductility recover of the material, what favors built-up edge and continuous chip formation;
- The tool size and heat treatment as well as its interaction presented a statistically significant influence on surface roughness results;
- Feed per tooth interaction with surface roughness results was inconclusive;
- In all experiments made, there is a predominance of peaks on the surface obtained, as *Ssk* values were greater than zero;
- As the *Sku* values were greater than three for all experiments, it can be concluded that the surface presents extreme peaks and valleys, what can be related to the chips left on the surface even after ultrasonic cleaning.

Acknowledgments. The authors thank the National Council for Scientific and Technological Development (CNPq) for the financial support, the LFS-POLI/USP and LEPU-UFU laboratories for all the support and equipment.

References

1. Baldo, D.: Study of micro milling of Ti-6Al-4V titanium alloy using signal analysis of force and acoustic emission. M.Sc. thesis, Federal University of São João del-Rei, São João del-Rei (2013)
2. Uhlmann, E., Kersting, R., Klein, T.B., Cruz, M.F., Borille, A.V.: Additive manufacturing of titanium alloy for aircraft components. In: Procedia CIRP, vol. 35, pp. 55–60 (2015). https://doi.org/10.1016/j.procir.2015.08.061
3. Chen, L.Y., et al.: Anisotropic response of Ti-6Al-4V alloy fabricated by 3D printing selective laser melting. Mater. Sci. Eng. A **682**, 389–395 (2017). https://doi.org/10.1016/j.msea.2016.11.061
4. Campos, F., Araujo, A.C., Munhoz, A.L.J., Kapoor, S.G.: The influence of additive manufacturing on the micromilling machinability of Ti6Al4V: a comparison of SLM and commercial workpieces. J. Manuf. Process. **60**, 299–307 (2020). https://doi.org/10.1016/j.jmapro.2020.10.006
5. Cardoso, P., Davim, J.P.: Optimization of surface roughness in micromilling. Mater. Manuf. Processes **25**(10), 1115–1119 (2010). https://doi.org/10.1080/10426914.2010.481002
6. Aksin, A., Karpat, Y.: Investigating microstructure effects of heat-treated commercially pure titanium (cp-Ti) based on mechanistic modeling of micro milling. In: 17th CIRP Conference on Modelling of Machining Operations, vol. 82, pp. 166–171 (2019)
7. Ahmadi, M., Karpat, Y., Acar, O., Kalay, Y.E.: Microstructure effects on process outputs in micro scale milling of heat treated Ti6Al4V titanium alloys. J. Mater. Process. Technol. **252**, 333–347 (2018). https://doi.org/10.1016/j.jmatprotec.2017.09.042
8. Attanasio, A., Gelfi, M., Pola, A., Ceretti, E., Giardini, C.: Influence of material microstructures in micromilling of Ti6Al4V alloy. Materials **6**(9), 4268–4283 (2013). https://doi.org/10.3390/ma6094268

9. Berglund, J., Soderberg, R., Warmefjord, K., Leach, R., Morse, E.: Functional tolerancing of surface texture – a review of existing methods. In: 16th CIRP Conference on Computer Aided Tolerancing (2020)
10. Etesami, S.A., Fotovvati, B., Asadi, E.: Heat treatment of Ti6Al4V alloy manufactured by laser-based powder bed fusion: process, microstructures, and mechanical properties correlations. J. Alloys Compounds **895**, Part 2 (2022). https://doi.org/10.1016/j.jallcom.2021.162618
11. Olympus: Surface roughness measurement - Parameters (2021). https://www.olympus-ims.com/en/metrology/surface-roughness-measurement-portal/parameters/
12. Seika, A.G., Kowalski, G.: Influence of roughness on the adhesion of coating on M2 high-speed steels with duplex treatment. Undergraduate diploma thesis, Federal Technological University of Paraná, Curitiba (2014)
13. Silva, J.: Correlation between surface texture and corrosion of PAPVD processed conjugates: monolayer and duplex Cr-N. M.Sc. thesis, Federal University of Minas Gerais, Belo Horizonte (2004)

Experimental Study of Micromilling of H13 Tool Steel Manufactured by Directed Energy Deposition

Milla Caroline Gomes[1](✉), Márcio Bacci da Silva[1], Wayne N. P. Hung[2], and Reginaldo Teixeira Coelho[3]

[1] Federal University of Uberlândia, Campus Santa Mônica, Uberlândia, MG, Brazil
millagomes@ufu.br
[2] Texas A&M University, College Station, TX, USA
[3] University of Sao Paulo, Sao Carlos, SP, Brazil

Abstract. Parts manufactured by additive manufacturing (AM) processes have rough surface finish and poor dimensional accuracy, therefore, post-processing must be required. Among the post-processing techniques, micro-milling is popular. Since machinability of materials manufactured by AM differs from those manufactured conventionally, this work investigates the micromilling of AISI H13 tool steel manufactured by the laser-beam directed energy deposition (LDED) and compares it with the micromilling of the same steel obtained by hot rolling and annealing heat treatment. Both quality of the machined surface and the cutting force were analyzed to assess the machinability. The force signals were acquired using a Kistler dynamometer and signal amplifier, and a National Instruments acquisition board and a computer with LabView Signal Express software. Roughness of a micromilled channel was measured using a Taylor Hobson profilometer. Micromilling was performed with the Minitech micromilling system with a maximum speed of 60,000 rpm. Mitsubishi's (Al,Ti)N coated carbide tool, with a cutting diameter of 400 μm were used. The cutting parameters included a cutting speed of 12.6 m/min, feed per tooth of 10 μm/tooth, axial depth of cut of 40 μm and radial depth of cut equal to 400 μm. Longer milling time caused a reduction of surface roughness Ra and an increase in the cutting force. No statistically significance differences between the Ra values and the cutting force obtained for micromilling the hot rolled versus LDED samples. However, the wear results of the microtools were compatible with the differences in the hardness of the analyzed samples.

Keywords: Micromilling · Additive Manufacturing · H13 Tool Steel

1 Introduction

Micromilling is an important micromachining process that has been widely used because it allows the manufacture of complex geometries with high precision on many materials (Serje et al., 2020). Due to these characteristics, micromilling has been used as a post-processing technique for materials manufactured by additive manufacturing, precisely

D. de Oliveira et al. (Eds.): COBEF 2023, LNME, pp. 183–189, 2024.
https://doi.org/10.1007/978-3-031-43555-3_17

because it can enhance both surface quality and geometric inaccuracies (Salonitis et al., 2016; Gomes et al., 2021).

In micromilling, the dimensions of the cutting tools vary from 1 μm to 1000 μm (Aramcharoen et al., 2008; Câmara et al., 2012). Rodrigues and Jasinevicius (2017), claimed that in this process, the dimensions involved are on the micrometric scale and defined the range of some of the process variables that correspond to cutting on a micro-scale. The values that have been set are:

I. Uncut chip thickness: 5 μm~20 μm;
II. Axial depth of cut: 1 μm~100 μm;
III. Tool edge radius: 1 μm~5 μm.

Therefore, it is noted that there is no consensus regarding the definition of the micromilling process, which is incomplete when considering only the values of the process variables or the diameter of the microtool. This difficulty is related to the specific phenomena that occur in micromachining, explained by the size effect. Therefore, it is not an easy task to establish a limit, from which the presence of the scale effect will have a greater influence (Simoneau et al., 2006). Thus, it is suggested that the concept to be given regarding micromilling should be a combination of the mentioned concepts since one would complement others.

The presence of the size effect is characterized by the significant increase in the specific energy, from the reduction of the uncut chip thickness (Liu and Melkote, 2007). This phenomenon explains the mechanisms that occur in micromilling, due to scale reduction in relation to macro-scale milling. This effect is present when the uncut chip thickness becomes comparable to the radius of the cutting edge and the grain of the machined material. Therefore, in micromilling, the grain size of the material, the uncut chip thickness, and the radius of the cutting edge have a great influence on the process (Cheng and Huo, 2013).

Additive manufacturing processes are being increasingly used as they enable the manufacture of complex geometries (Zhai et al., 2014). However, parts manufactured by these processes usually have a low surface finish and dimensional accuracy. Thus, after manufacturing these parts, they are subjected to post-processing techniques, such as micromilling, with the aim of improving surface quality and dimensional accuracy (Kumbhar and Mulay, 2018).

Of the additive manufacturing processes, those used in the manufacture of metallic parts are binder jetting (BJ), sheet lamination (SL), directed energy deposition (DED), and powder bed fusion (PBF) among which the last two are mostly used (Zhang et al., 2018). The popular DED is developed for different metals, and repairing damaged or worn parts (Yi et al., 2019).

In general, the machinability of AM metals differs from the same metals manufactured by conventional methods, with metals obtained by additive manufacturing having lower machinability (Hung and Corliss, 2019). Therefore, it becomes important to investigate the main differences between micromilling of conventional materials and those manufactured by additive manufacturing. This work aims to carry out an experimental study to investigate the micromilling of AISI H13 tool steel manufactured by hot rolling and by DED.

2 Methodology

The workpiece used in carrying out the tests was AISI (American Iron and Steel Institute) H13 tool steel manufactured by LDED. The H13 tool steel is commonly used for high temperature die and mold as in hot forming or injection molding (Asnafi, 2021).

The chemical composition in the percentage of the weight of the two analyzed samples is shown in Table 1. The composition of the hot rolling sample was obtained by the Materials Testing and Analysis Laboratory – LAMAT of SENAI Itaúna, while that of the LDED sample was obtained by EDS using the SEM from Hitachi High-Technologies®, TM3000. Note that the difference in the amount of carbon is due to the inaccuracy of the SEM system in detecting this chemical element.

Table 1. Chemical composition of AISI H13 tool steel (LAMAT and EDS)

H13	Chemical composition (wt%)										
	C	Si	Mn	P	S	Cr	Mo	Al	Cu	Ni	V
Hot Rolled	0.380	0.910	0.300	0.010	0.005	5.090	1.270	0.010	0.080	0.310	0.850
LDED	8.155	1.128	0.431	0.081	–	5.286	1.049	0.019	0.004	–	0.745

Table 2 shows the hardness of each analyzed sample.

Table 2. AISI H13 tool steel hardness

H13	Hardness (HV)
Hot Roled	204
LDED	513

The micromilling tests, for the manufacture of 10 microchannels in each analyzed 13 × 12 × 5 mm sample, were carried out on a CNC Mini-mill/GX model micromilling machine, manufactured by Minitech Machinery Corporation®, which has a maximum spindle rotation of 60,000 rpm, 3 axes with a resolution of the positioning of 0.1 μm. It was installed on an inertial table, specially designed to reduce the influence of vibrations from external sources during the tests.

Blaser's Vascomill MMS FA 2 cutting fluid was used in a minimum quantity fluid (MQF) regime at a pressure of 6 bar. The positioning of the fluid nozzle was at an approximate distance of 10 mm from the microtool and inclination of 20°, as can be seen in Fig. 1.

The tools used were micromills manufactured by Mitsubishi Materials of carbide coated with (Al, Ti)N. The tool geometry included two cutting flutes cutting diameter (DC) of 0.4 mm, cutting length (APMX) of 0.8 mm, diameter (DCON) and length (LF)

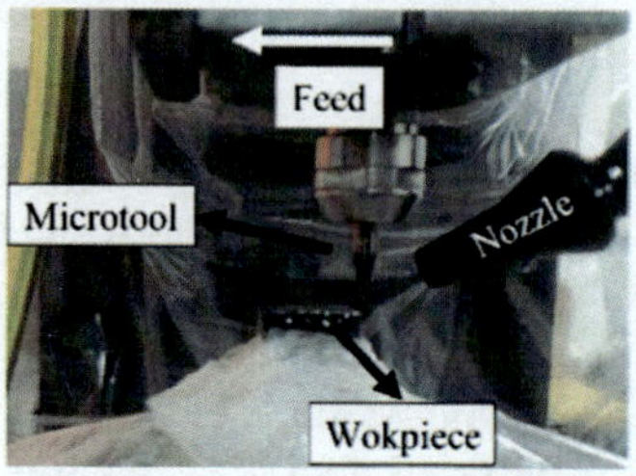

Fig. 1. Experimental setup.

of 4 mm and 40 mm, respectively. Figure 2 illustrates the microtools and identification of each geometric parameter.

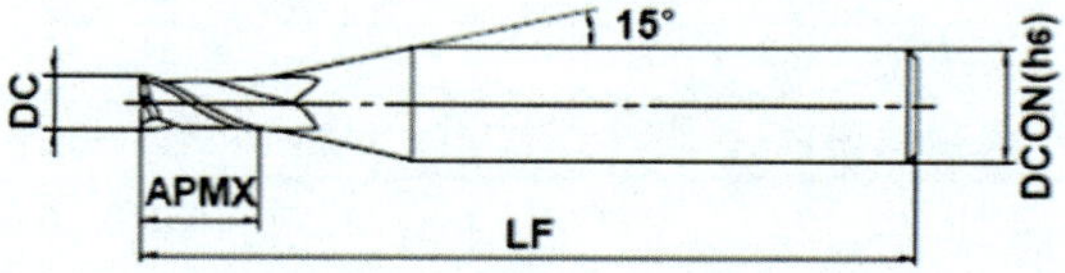

Fig. 2. Geometric parameters of microtools (Mitsubishi, 2021).

The cutting condition used in the tests is shown in Table 3.

Table 3. Cutting parameters

Cutting Speed (m/min)	Feed per tooth (μm/tooth)	Feed Rate (mm/min)	Axial Depth of Cut (μm)
12.56	10	100	40

The Taylor Hobson® profilometer, model Form Talysurf Intra, was used to measure the surface roughness. The probe scanned near center and along a micromilled channel. Three measurements were obtained in different regions of each microchannel, using a Gaussian filter, cut-off of 0.8 mm, and evaluation length of 4 mm, according to ISO 4288 (1996). For analysis, the average of these measurements and the standard deviation were calculated. The parameter analyzed was the mean arithmetic deviation (Ra).

During the fabrication of each microchannel, the machining force in x-direction was acquired with a data acquisition system. Therefore, the machining force was acquired in the direction perpendicular to the cutting direction, since the objective was to analyze the highest value of the cutting force, which corresponds to the peaks of the acquired force, corresponding to the force obtained in the maximum uncut chip thickness (hcu), as represented in Fig. 3.

The Kistler 9265B dynamometer with a 9443B mounting plate was used to secure and mounted a machined sample. This dynamometer was connected to a Kistler 5407

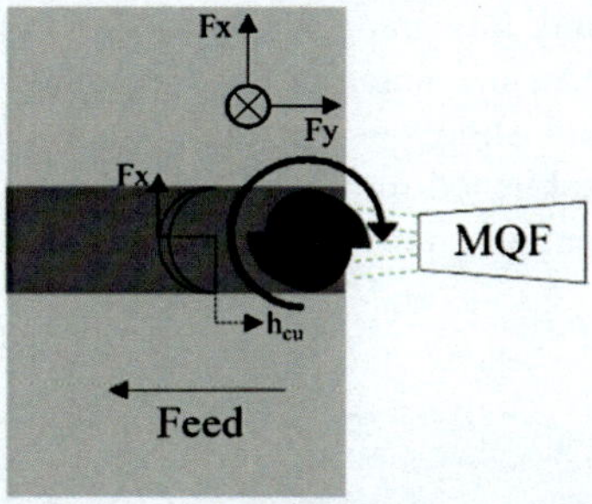

Fig. 3. Representation of the acquired cutting force.

distribution box, which in turn connected to the Kistler 504E load amplifier. From the amplifier, the signals were sent to the National Instruments NI USB-62211 acquisition board to carry out the conversion from an analog to a digital signal. Finally, the LabView Signal Express software was used to read the digital signal. Figure 4 shows how this signal acquisition system was assembled.

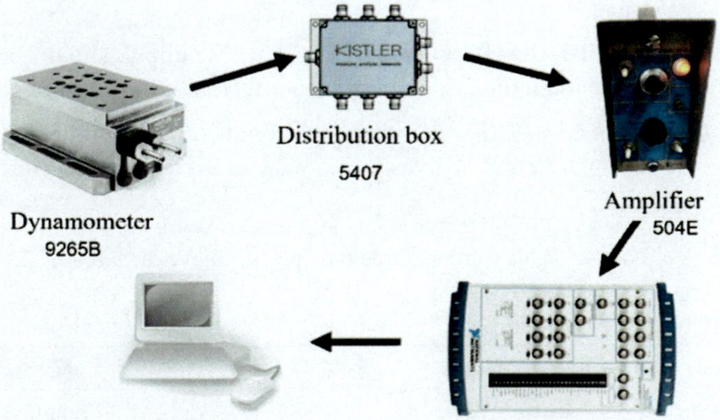

Fig. 4. Analyzed force acquisition system.

In the analysis of the acquired force, the mean force of the peaks of the signals collected during the machining period was used, since the peak of the signals of the Fx component corresponds to the highest value of the cutting force for the maximum uncut chip thickness. Matlab was the software used to process the signals. The acquisition rate was 10 kHz.

To measure the wear of the microtool when machining the last microchannel, the microtools were observed in the SEM, using the Scanning Electron Microscope (SEM), from Hitachi High-Technologies model TM3000, for analysis and obtaining images. It is important to emphasize that images of the new microtools were also obtained to verify the existence of any defects in them. The wear was obtained by analyzing the images of the secondary flank face images (top images) of the micromills and for its measurement the images of the new microtool were superimposed on the images of the used ones, using Adobe Photoshop CC software. Then, the wear was measured with Image J software, being established for each of the two cutting flutes, as the distance between the tip of the new and worn tool cutting edge, as schematized in Fig. 5.

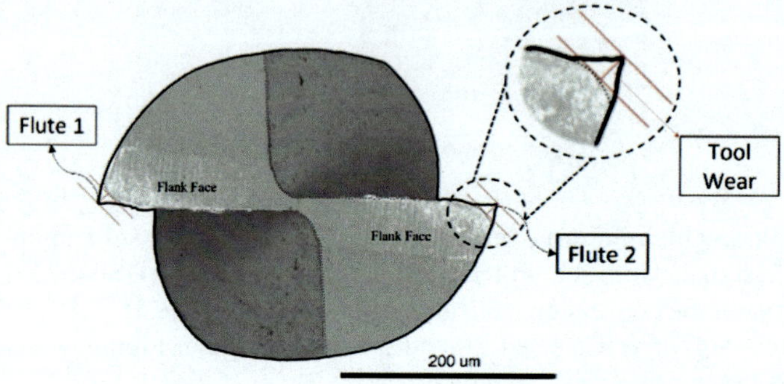

Fig. 5. Tool wear measurement (Gomes et al., 2020).

Sample Heading (Third Level). Only two levels of headings should be numbered. Lower level headings remain unnumbered; they are formatted as run-in headings.

Sample Heading (Forth Level). The contribution should contain no more than four levels of headings. The following Table 4 gives a summary of all heading levels.

Table 4. Table captions should be placed above the tables.

Heading level	Example	Font size and style
Title (centered)	**Lecture Notes**	14 point, bold
1st-level heading	**1 Introduction**	12 point, bold
2nd-level heading	**2.1 Printing Area**	10 point, bold
3rd-level heading	**Run-in Heading in Bold.** Text follows	10 point, bold
4th-level heading	*Lowest Level Heading.* Text follows	10 point, italic

Displayed equations are centered and set on a separate line.

$$x + y = z \quad (1)$$

Please try to avoid rasterized images for line-art diagrams and schemas. Whenever possible, use vector graphics instead (see Fig. 6).

For citations of references, we prefer the use of square brackets and consecutive numbers. Citations using labels or the author/year convention are also acceptable. The following bibliography provides a sample reference list with entries for journal articles [1], an LNCS chapter [2], a book [3], proceedings without editors [4], as well as a URL [5].

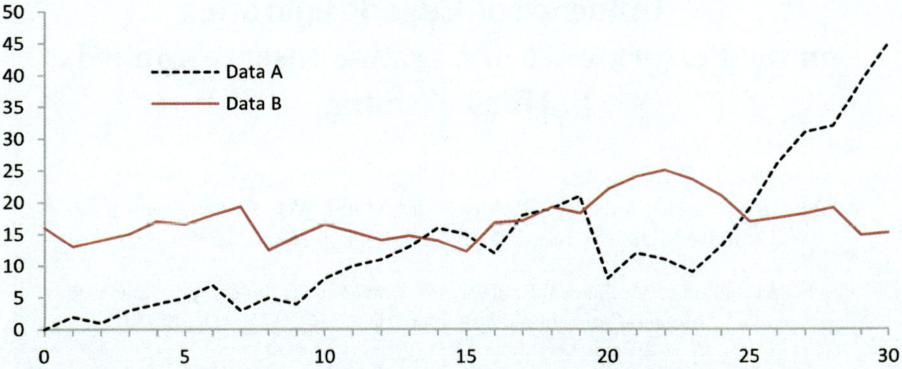

Fig. 6. A figure caption is always placed below the illustration. Short captions are centered, while long ones are justified. The macro button chooses the correct format automatically.

References

1. Author, F.: Article title. Journal **2**(5), 99–110 (2016)
2. Author, F., Author, S.: Title of a proceedings paper. In: Editor, F., Editor, S. (eds.) CONFERENCE 2016, LNCS, vol. 9999, pp. 1–13. Springer, Heidelberg (2016)
3. Author, F., Author, S., Author, T.: Book title, 2nd edn. Publisher, Location (1999)
4. Author, F.: Contribution title. In: 9th International Proceedings on Proceedings, pp. 1–2. Publisher, Location (2010)
5. LNCS Homepage. http://www.springer.com/lncs. Accessed 21 Nov 2016

The Influence of Edge Preparation on the Performance of Ceramic Inserts Applied to Hard Turning

Augusto Moura Martins[1(✉)], Alexandre Mares Weikert de Oliveira[2], Alexandre Mendes Abrão[1], and Carlos Eiji Hirata Ventura[3]

[1] Graduate Program in Mechanical Engineering, Universidade Federal de Minas Gerais, Avenida Antônio Carlos 6627, Belo Horizonte, MG 31270-901, Brazil
augusto.mmartins2@gmail.com

[2] Mechanical Engineering Undergraduate Program, Universidade Federal de Minas Gerais, Avenida Antônio Carlos 6627, Belo Horizonte, MG 31270-901, Brazil

[3] Graduate Program in Mechanical Engineering, Universidade Federal de São Carlos, Rodovia Washington Luís km 235, São Carlos, SP 13565-905, Brazil

Abstract. The need for mechanical components with superior strength and specific characteristics for application under severe conditions requires materials with satisfactory mechanical strength and wear resistance under elevated loads. Hardened steels present suitable properties for these applications, thus, manufacturing processes must be capable of dealing with these characteristics to produce high quality components. The aim of this work is to investigate the influence of the cutting conditions (cutting speed and feed) and cutting edge preparation (chamfer angle) of mixed alumina tools (Al_2O_3 + TiC) when turning hardened AISI 52100 bearing steel (62 HRC), with emphasis on the turning force components, machined surface finish and tool life. The results indicated that feed statistically affected all the evaluated parameters (increasing cutting, feed and passive forces and roughness), whereas cutting edge preparation affected only the feed force and surface roughness (lower feed force and roughness obtained using the tool with a chamfer 80 μm wide and angle of 30º). Regarding tool life, the commercial tool and the tool prepared with chamfer width of 80 μm and angle of 30º presented superior performance and allowed gradual tool wear evolution, in contrast with the other cutting edge preparations which led to edge fracture.

Keywords: Hard Turning · Ceramic Tool · Cutting Edge Preparation

1 Introduction

Turning is one of the most relevant machining operations, being widely used for the production of rotational parts with various dimensions and geometries, distinct materials and levels of hardness. In addition to roughing, hard turning can be used for finishing as an alternative to abrasive processes, thus reducing the operation time and cost at the expense of wider tolerances. When turning is performed on materials with hardness

D. de Oliveira et al. (Eds.): COBEF 2023, LNME, pp. 190–200, 2024.
https://doi.org/10.1007/978-3-031-43555-3_18

between 58 and 68 HRC, it is known as hard turning and is typically applied to hardened steels [1]. Hard turning requires the use of machine tools with elevated stiffness, stability and precision to provide tight tolerances and low surface roughness. To achieve these goals, cutting tools with elevated hardness and wear resistance are required. Most of the hard turning operations occurs without the application of cutting fluids, as the cutting tool temperature causes its instantaneous ebullition, indicative condition of the higher severity of the operation [2].

During hard machining, chamfered cutting edges are preferably used. Round edges are not commonly employed because they limit tool life and promote unstable cutting conditions. The number of chamfers can vary, as their width and angle. The rake and flank surfaces can also present an inclination, depending on the characteristics of the material to be machined. The commonly used values for chamfer width of turning tools applied to hard turning are between 100 and 200 μm with chamfer angle ranging between 10° and 35°. Chamfering is generally produced on alumina-based ceramic and polycrystalline cubic boron nitride (PcBN) cutting tools [3]. Cutting edge preparation modifies the cutting wedge geometry, affecting the deformation process of the work material, as well as energy consumption and turning forces. Kim *et al.* [4] compared the influence of tool material (PcBN and alumina-based ceramic) and cooling environment (dry, flooding and cryogenic cooling with liquid nitrogen) on the behavior of hard turning of AISI 52100 bearing steel (62 HRC) and noticed that cryogenic cooling had a positive effect on the reduction of energy and time consumption, machining cost, tool life and environmental impact, irrespectively of the cutting tool grade. In contrast, dry cutting with ceramics represented the poorest overall performance, exception made for machining cost.

Regardless of its hardness and wear resistance, the cutting tools will be subject to wear mechanisms and eventually its replacement will be necessary. When hard turning, the high normal compressive and shear stresses associated with high temperatures in the cutting zone drastically impair the performance of cutting tools with inferior properties or inappropriate geometry. On the other hand, proper edge preparation can induce compressive residual stresses of higher intensity on the machined surface and prolong the service life of the component. According to Hua *et al.* [5], the larger the edge radius or wider the chamfer, the deeper and more compressive residual stresses are induced on a hardened steel.

Tool wear rate and evolution are considerably affected by cutting edge preparation. Agmell *et al.* [6] reported that chamfered tools promoted a larger stagnation zone ahead of the cutting tool in comparison with rounded edges. In contrast, Zhou *et al.* [7] investigated the influence of the chamfer angle on the wear of PcBN tools and noticed a reduction of up to 53% in the wear land of chamfered tools. Özel [8] compared the performance of PcBN tools with constant edge radii and tools with variable edge radii and reported that the latter promoted best results in terms of heat generation, stress concentration and tool life.

The use of cutting tools with different edge preparations can significantly affect the performance of hard turning. Therefore, the present work intends to evaluate the influence of edge geometry and turning parameters (cutting speed and feed) on the turning force components, machined surface roughness and tool wear when turning hardened AISI 52100 bearing steel (62 HRC).

2 Materials and Methods

AISI 52100 bearing steel (0.98 – 1.10% C; 0.15 – 0.35% Si; 0.25 – 0.45% Mn e 1.30 – 1.60% Cr) was selected as wok material. Tubes with an external diameter of 65 mm, internal diameter of 38 mm and length of 300 mm were used. The work material was normalized and further quenched and tempered, thus resulting in an average hardness of 62.5 ± 1 HRC.

Edge preparation was undertaken on an Agathon DOM Plus grinding machine (maximum power of 16 kW and maximum rotational speed of 3400 rpm, Fig. 1a). A diamond griding wheel with D15 grain size and vitrified bond with C100 concentration was used, together with the following parameters: cutting speed $v_c = 30$ m/s, axial feed velocity $v_{fa} = 1$ mm/min and insert rotation of $v_R = 5{,}75°/s$. Before griding each insert, the wheel was dressed with an Al_2O_3 abrasive, 220 grit, dressing roll with a dressing cutting speed $v_{cd} = 10$ m/s, axial speed $v_{fad} = 0{,}3$ mm/min and dressing length $l_d = 30$ μm.

The turning tests were conducted on a computer numerically controlled lathe (maximum power of 5.5 kW and maximum rotational speed of 3,500 rpm) using mixed alumina cutting tools (Sandvik grade CC650) with ISO geometry SNGN 120408 T01020 mounted on a tool holder code MSSNR 2020 K12. In addition to the commercial edge preparation, distinct chamfer geometries and sharp inserts were produced and tested. The edge preparation parameters and cutting conditions are given in Table 1. The turning tests were conducted at random, following a full factorial design with two replicates (60 runs). Type C edge preparation presents a chamfering identical to the commercial tool and the sharp tool presents an edge radius of 10.34 μm.

Table 1. Experimental conditions.

Parameters	Values				
Edge preparation	A $\gamma = 10°$ $b = 80$ μm	B $\gamma = 30°$ $b = 80$ μm	C $\gamma = 20°$ $b = 80$ μm	Commercial $\gamma = 20°$ b = 80 μm	Sharp $\gamma = 0°$ $b = 0$ μm
Cutting speed v_c [m/min]	60	90	120		
Feed f [mm/rev]	0.08	0.16			
Depth of cut [mm]	1				

Figure 1b presents the experimental setup on the CNC lathe together with the cutting edge characteristics (chamfer width b and chamfer angle γ) schematically. The tool holder was mounted on a Kistler 9272 dynamometer to allow the measurement of the components of the turning force. Machined surface roughness (arithmetical mean deviation - Ra) was assessed using a Taylor Hobson Surtronic 25 surface roughness tester. The statistical analysis (significance level of 0.05) of the results was performed using Minitab 15 software.

The tool life tests were performed at a cutting speed of 60 m/min, feed of 0.08 mm/rev and depth of cut of 1.0 mm. After 10 min of continuous cutting, flank wear was measured using an Olympus CX31 optical microscope. Three dimension images of fresh and worn cutting tools were obtained with an Alicona InfiniteFocus SL confocal microscope (Fig. 1c).

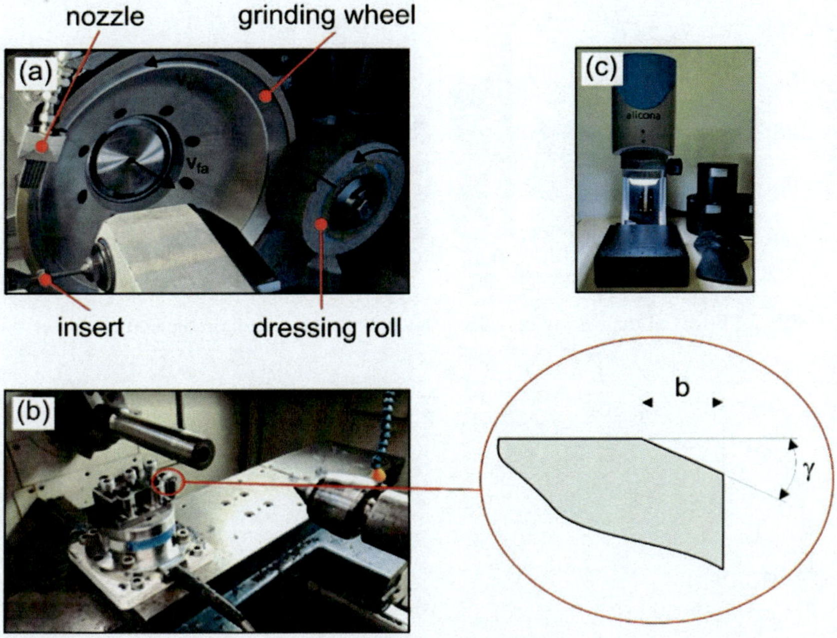

Fig. 1. Experimental setup: (a) grinding machine, (b) dynamometer mounted on the lathe and chamfer geometry (detail), (c) confocal microscope.

3 Results and Discussion

Firstly, the results concerned with the cutting, feed and thrust forces are presented, followed by roughness of the machined surface and tool life/wear.

3.1 Turning Force Components

The influence of the cutting conditions and edge preparation on the cutting force can be seen in Fig. 2. In general, it can be noticed that the cutting force decreases with the elevation of cutting speed (owing to the reduction in the shear strength of the work material caused by higher temperatures) and increases with feed (elevation of the shear area). However, a clear trend is not evident with regard to the influence of edge preparation. As a matter of fact, the Pareto chart presented in Fig. 3 indicates that feed is the only factor that significantly affects the cutting force and Fig. 4 shows that a two-fold increase in feed does not result in a proportional increase in the cutting force, probably because when the highest feed (0.16 mm/rev) is used, chip is partly formed on the chamfer and partly on the rake face and this effect changes the material deformation characteristics.

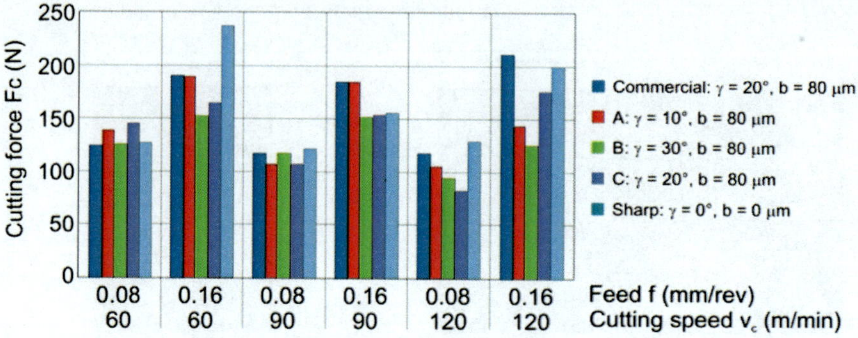

Fig. 2. Effect of the cutting conditions and edge preparation on the cutting force.

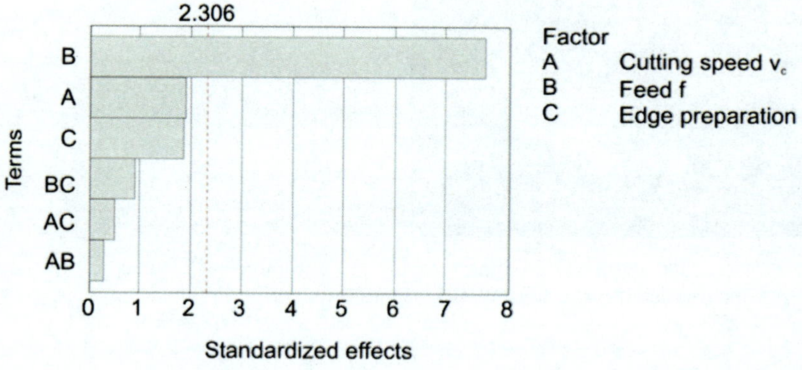

Fig. 3. Pareto chart for the cutting force.

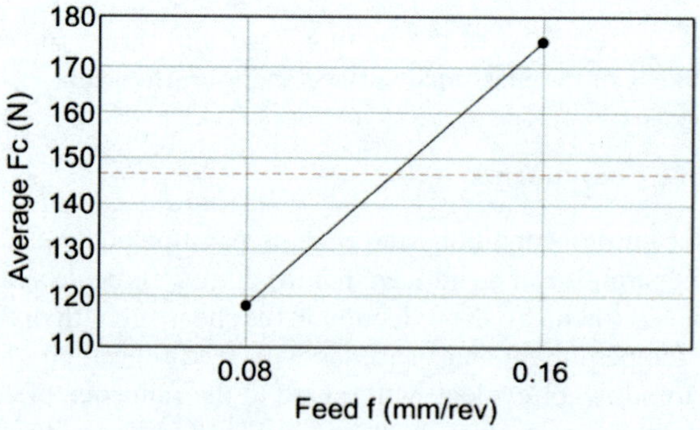

Fig. 4. Main effect (feed) plot for cutting force.

Figure 5 shows the influence of cutting speed, feed and edge preparation on the feed force. Not surprisingly, the feed force increased with feed for the same reason previously explained (increase in the shear area). Cutting speed seems to slightly affect the feed force, as well as edge preparation (in general, lower feed forces were recorded for the commercial tool and edge preparation B, i.e. $\gamma = 30°$ b $= 80$ μm, while the sharp tool and edge preparation A ($\gamma = 10°$ b $= 80$ μm) produced higher feed force, especially when combined with the highest feed.

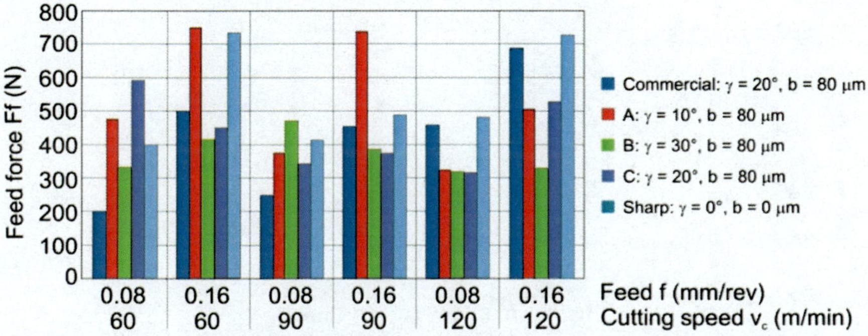

Fig. 5. Effect of the cutting conditions and edge preparation on the feed force.

The Pareto chart presented in Fig. 6 indicates that feed and edge preparation were statistically significant for feed force and their effect can be observed in Fig. 7. Similarly to the cutting force, feed force increases with feed due to the increase in the shear area. With regard to the influence of edge preparation, edge preparation B ($\gamma = 30°$ b $=$ 80 μm) promoted the lowest feed force values and the sharp tool the highest. Although this preparation represents the steepest negative chamfer angle, this feature did not reflect on the feed force, probably because the chamfer width is equal to the lower feed value used. Therefore, the extent of the contact between the chip and the tool rake face will prolong beyond the chamfer width, reducing the influence of the chamfer angle. Edge

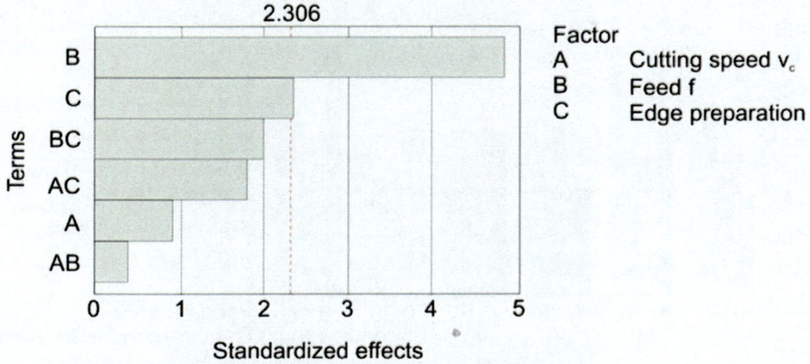

Fig. 6. Pareto chart for the feed force.

preparation has a crucial influence on the feed force, since the latter represents the reaction of the work material against the feed movement (in contrast to the cutting force, that acts on the rake face). In addition to that, the commercial and C tools presented slight differences on the feed force values, an unexpected result since both tools possess the same geometry. A possible explanation for this behavior can be attributed to the edge preparation conditions (griding wheel and parameters and dressing conditions). These differences can result in distinct topographies on the chamfers that would affect chip seizure and sliding conditions during turning.

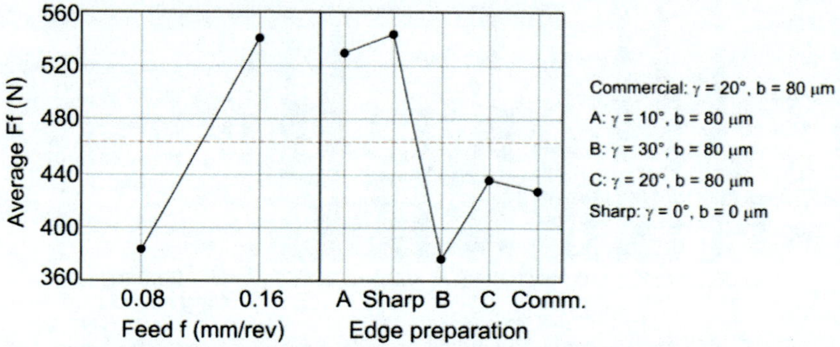

Fig. 7. Main effect plots (feed and edge preparation) for feed force.

The influence of the cutting conditions and edge preparation on the behavior of the thrust force is presented on Fig. 8. Neither the cutting conditions nor the edge preparation presented an evident influence on this component of the turning force and the Pareto chart indicated that one of the investigated factors was statistically significant. In contrast to the cutting and feed components of the turning force, the thrust force is not associated with any motion and simply represents the resistance of the work material against the depth of cut, which was kept constant throughout the experimental program. Considering the depth of cut value used ($a_p = 1.0$ mm) and that the cutting tool possesses a nose radius of 0.8 mm and cutting edge angle of 45°, the area of contact normal to the thrust

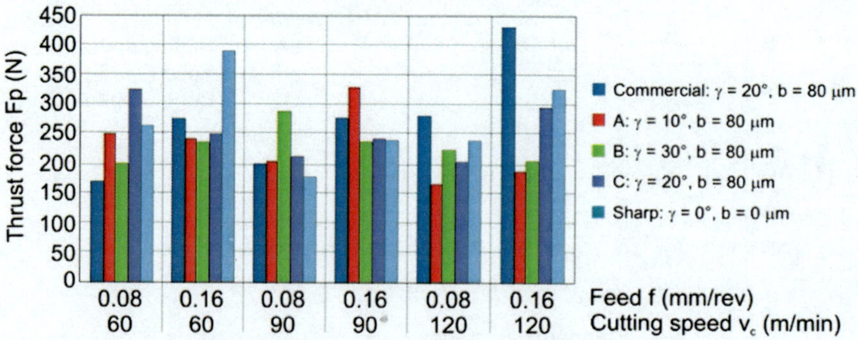

Fig. 8. Effect of the cutting conditions and edge preparation on the thrust force.

force is hardly altered by cutting speed, feed or edge preparation, therefore, thrust force is not drastically affected.

3.2 Surface Roughness

Figure 9 shows the effect of the cutting conditions and edge preparation on the arithmetical mean deviation (Ra) of the turned surfaces. As expected, increasing feed has a clear influence on surface roughness, while the influence of cutting speed is not evident. As far as the influence of edge preparation is concerned, the tools with preparations B ($\gamma = 30°$, $b = 80\ \mu m$) and C ($\gamma = 20°$, $b = 80\ \mu m$) provided the best results, followed by the commercial tool in some cases. Moreover, the sharp tool presented the overall worst performance, probably due to chipping during the tests, and was not able to withstand $v_c = 120$ m/min and $f = 0.16$ mm/rev for a cutting length that allowed the measurement of Ra. The Pareto chart on Fig. 10 indicates that feed and edge preparation are the only statistically significant factors and the main effects plot in Fig. 11 shows that the tools with edge preparation B provides the lowest roughness values, while the sharp edge promotes the worst results. Interestingly, such behavior is similar to that observed for the feed force, thus suggesting that the improve in surface finish is directly affected by edge preparation. In addition to the increase in the shear area with the elevation of feed (which induces higher plastic deformation and requires higher forces, thus impairing the surface finish), the elevation of feed contributes to the increase in the distance between roughness peaks and valleys. Finally, the differences in the Ra values promoted by the commercial tool and that with edge preparation C are attributed to the distinct grinding conditions employed in the preparation of each tool.

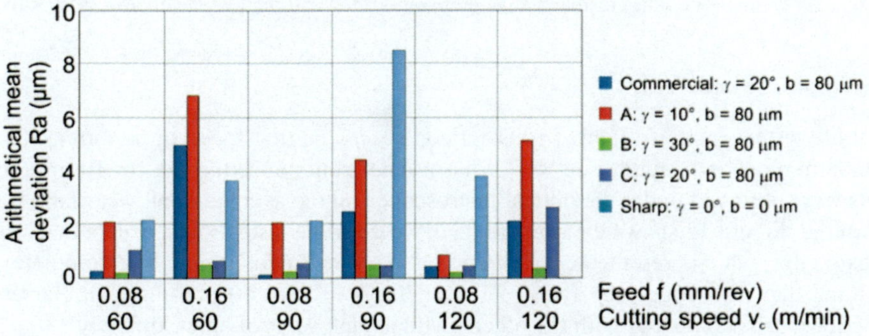

Fig. 9. Effect of the cutting conditions and edge preparation on the arithmetical mean deviation (Ra)

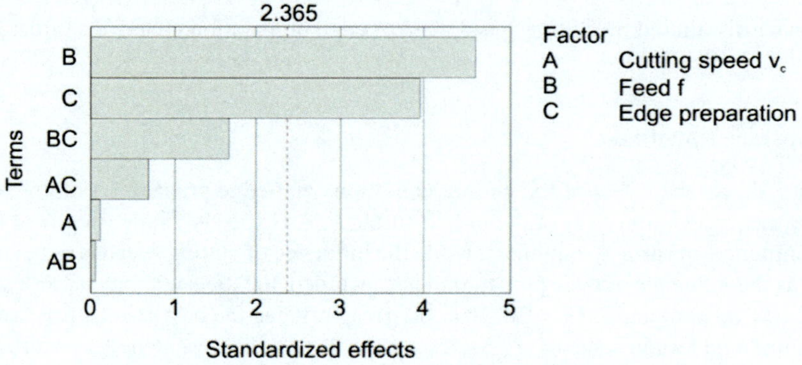

Fig. 10. Pareto chart for the arithmetical mean deviation (Ra).

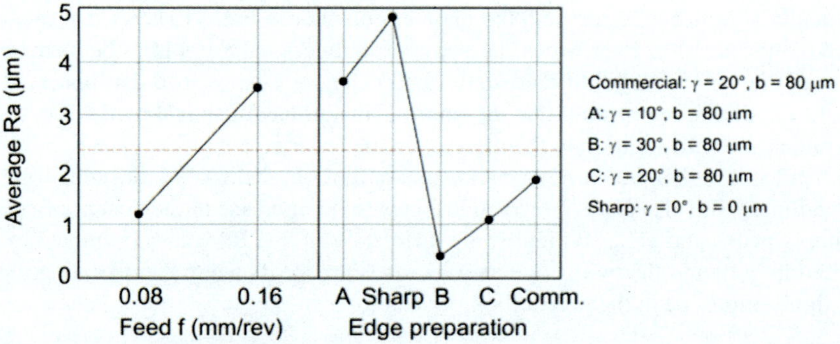

Fig. 11. Main effect plots (feed and edge preparation) for arithmetical mean deviation (Ra).

3.3 Tool Life and Wear

Tool life tests were performed at constant conditions (cutting speed of 60 m/min, feed of 0.08 mm/rev and depth of cut of 1.0 mm). After continuous turning for 10 min, the tools were observed under the optical microscope and the average flank wear land was measured. Figure 12 shows the average flank wear for the tested tools, except that with a sharp edge, which presented catastrophic failure after 3 min cutting. The commercial tool and that with edge preparation B ($\gamma = 30°$, $b = 80$ µm) presented similar and superior results compared with the others, with a flank wear of approximately 180 µm after the end of the test. On the other hand, edge preparation C ($\gamma = 20°$, $b = 80$ µm), identical to the commercial, presented elevated wear, for the same reason previously given: distinct conditions used in chamfer grinding.

Figures 13a and 13b present, respectively, the images of the cutting edges generated by the focus-variation microscope before and after the tool life tests. As indicated in Fig. 12, the commercial tool and tool with preparation B presented the lowest flank wear, apparently caused by abrasion and microchipping. The tools with edge preparations A and C also presented abrasive wear, however, cutting edge fracture caused by chipping and spalling. On its turn, the tool with sharp edge was not capable of withstanding the stresses imposed during hard turning and presented catastrophic failure after 3 min only. Finally, considering all the evaluated parameters, it can be stated that the tool with edge preparation B ($\gamma = 30°$ b = 80 μm) provides best overall performance, i.e., it is capable of providing lower forces and roughness, in addition to presenting lower flank wear.

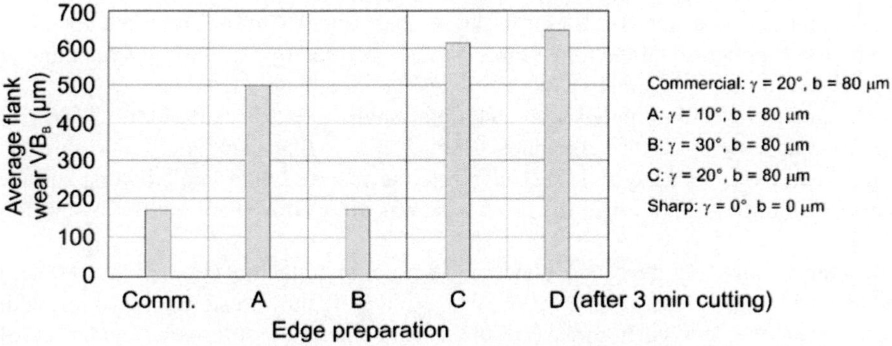

Fig. 12. Average flank wear for distinct edge preparations after 10 min using a cutting speed of 60 m/min, feed of 0.08 mm/rev and depth of cut of 1.0 mm.

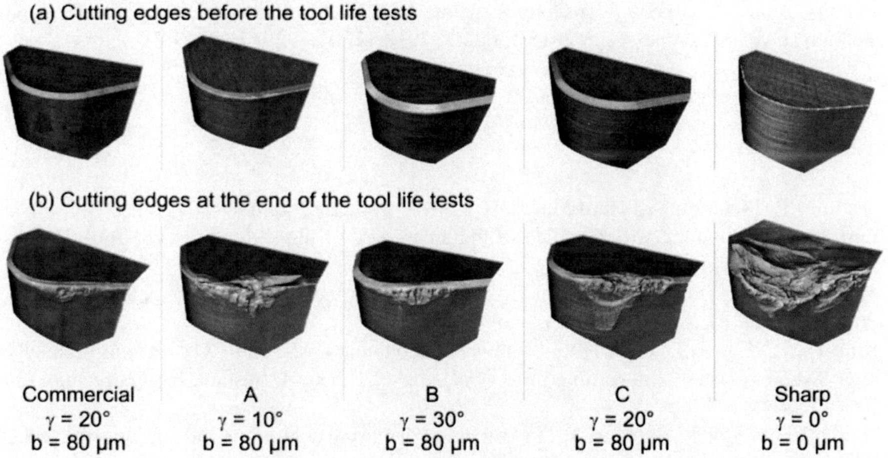

Fig. 13. Cutting tools with distinct edge preparations: (a) before and (b) after the tool life tests (cutting speed of 60 m/min, feed of 0.08 mm/rev and depth of cut of 1.0 mm.).

4 Conclusion

The following conclusions can be drawn from this study:

With regard to the turning force components, the cutting and feed forces were statistically affected by feed, however, the feed force was also influenced by edge preparation (the lowest feed force for the tool with chamfer angle $\gamma = 30°$ and width b = 80 μm and the highest force recorded for the tool with sharp edge). Cutting speed did not affect the force components and the thrust force was not affected by the cutting parameters or edge preparation.

Similarly, the tool prepared with a chamfer angle $\gamma = 30°$ and width b = 80 μm produced the lowest surface roughness (highest Ra = 0.54 μm), outperforming the commercial tool under the highest feed, while the sharp edge promoted worst surface finish. Within the tested range, the roughness was affected by feed but not by cutting speed.

As far as tool life and wear are concerned, the commercial tool ($\gamma = 20°$ b = 80 μm) and that with edge preparation B ($\gamma = 30°$ b = 80 μm) showed similar behavior, with longest lives and gradual wear, while the other preparations presented evidences of chipping and fracture, notably the sharp edge tool.

Acknowledgments. The authors are grateful to the National Council for Scientific and Technological Development (CNPq), to the Coordination for the Improvement of Higher Education Personnel—Brazil (CAPES) under Grant Agreement Code 001, and to the São Paulo Research Foundation (FAPESP, Processes nº 2015/15622-2, 2017/12304-5, 2017/12309-7).

References

1. Davim, J.P.: Machining of Hard Materials, 1st edn. Springer, London (2011)
2. Bartarya, G., Choudhury, S.K.: State of the art in hard turning. Int. J. Mach. Tools Manuf. **53**(1), 1–14 (2012)
3. Ventura, C.E.H., Köhler, J., Denkena, B.: Strategies for grinding of chamfers in cutting inserts. Precis. Eng. **38**(4), 749–758 (2014)
4. Kim, D.M., Kim, H.I., Park, H.W.: Tool wear, economic costs, and CO_2 emissions analysis in cryogenic assisted hard-turning process of AISI 52100 steel. Sustain. Mater. Technol. **30**, e00349 (2021)
5. Hua, J., Umbrello, D., Shivpuri, R.: Investigation of cutting conditions and cutting edge preparations for enhanced compressive subsurface residual stress in the hard turning of bearing steel. J. Mater. Process. Technol. **171**(2), 180–187 (2006)
6. Agmell, M., Ahadi, A., Gutnichenko, O., Stahl, J.-E.: The influence of tool micro-geometry on stress distribution in turning operations of AISI 4140 by FE analysis. Int. J. Adv. Manuf. Technol. **89**, 3109–3122 (2017)
7. Zhou, J.M., Walter, H., Andersson, M., Stahl, J.E.: Effect of chamfer angle on wear of PCBN cutting tool. Int. J. Mach. Tools Manuf **43**(3), 301–305 (2002)
8. Özel, T.: Computational modelling of 3D turning: Influence of edge micro-geometry on forces, stresses, friction and tool wear in PcBN tooling. J. Mater. Process. Technol. **209**(11), 5167–5177 (2009)

The Influence of Base Fluid and Graphene Nanoparticles Concentration on Surface Integrity of SAE 52100 Steel After Grinding

Raphael Lima de Paiva[1(✉)], Déborah de Oliveira[2], and Rosemar Batista da Silva[3]

[1] Federal University of Piauí, Teresina, PI, Brazil
raphaellimap@ufpi.edu.br
[2] University of Brasília, Brasília, DF, Brazil
[3] Federal University of Uberlândia, Uberlândia, MG, Brazil

Abstract. Grinding is known as a high specific energy process, in which most of it is transformed into heat and conducted to the workpiece. Thus, the use of cutting fluid is indispensable to attenuate or avoid the occurrence of thermal damage. However, concerning human health and the environment, as well as economic aspects and the search for more sustainable processes, strategies that reduce cutting fluid usage such as the minimum quantity lubrication (MQL) technique have gained prominence. This technique uses a low quantity of cutting fluid, which is directed to the cutting zone with compressed air. Nevertheless, although the MQL technique has shown promising results in comparison to dry grinding and flood technique for different materials, the use of nanofluids has been showing great potential to improve the MQL technique, since the presence of nanoparticles tends to improve lubrication and cooling capacities of the base fluid. In this sense, this work aims to evaluate the influence of the type of base fluid and nanoparticles concentration on the surface integrity of SAE 52100 hardened steel after grinding with graphene-based nanofluids applied with the MQL technique. Two different base fluids were tested: a semi-synthetic vegetable base and a synthetic one. The graphene was added to the base fluids at two different concentrations: 0.025 wt.% and 0.075 wt.%. Grinding trials using the base fluids only (without nanoparticles) were also performed for comparison purposes. The surface integrity of the workpiece after grinding was analyzed in terms of surface finish and microhardness below the machined surface. The results showed that the nanofluid's efficiency in reducing the surface roughness of the ground surface was strongly influenced by the base fluid and graphene concentration: best results were found after grinding with a combination between semi-synthetic fluid and the lowest graphene concentration (0.025 wt.%). Additionally, the presence of graphene in the cutting fluid attenuated the occurrence of thermal damage in terms of microhardness reduction; for the highest nanoparticles concentration (0.075 wt.%), such reductions were only 1% and 3% after grinding with semi-synthetic and synthetic fluid, respectively.

Keywords: Grinding · Nanofluids · Graphene · MQL · Surface integrity · Thermal damage

D. de Oliveira et al. (Eds.): COBEF 2023, LNME, pp. 201–209, 2024.
https://doi.org/10.1007/978-3-031-43555-3_19

1 Introduction

Due to some inherent characteristics of grinding (e.g., high cutting speeds, low radial depth of cut values that means low material removal rate, intense friction, plastic deformation in cutting mechanism, and high specific energy compared to machining processes with multipoint cutting tools with defined geometry, such as milling), the use of cutting fluid becomes indispensable to prevent the surface integrity of the workpiece from surface finishing deterioration and thermal damages as well as cutting efficiency of the grinding wheel. This is particularly important when grinding high-carbon hardened steels such as SAE 52100 due to the high prone for microstructure alteration under the elevated temperatures developed during the cutting process, especially considering conventional abrasives such as aluminum oxide (Al_2O_3) [1, 2].

However, the use of cutting fluid goes against the recent politics regarding sustainable and more environmentally friendly processes, as most of the cutting fluids, especially the mineral ones, are harmful to both human health and the environment, not to mention the costs for purchase, maintenance, and disposal [3, 4]. In this context, there is a need for cooling-lubrication strategies that can reduce and amount of cutting fluids, especially in grinding where flow rates are generally higher than 10 L/min (600,000 mL/h). So, the minimum quantity lubrication (MQL) technique is the most common and widely used one. It consists of delivering an extremely low quantity of fluid to the cutting zone (flow rates usually lower than 500 mL/h) with the aid of compressed air [5].

Despite some promising results in grinding hardened steels with the MQL technique, its poor cooling capacity when grinding some hardened materials is a concern and usually limits its widespread application in the metalworking industry, among others. In this sense, the addition of nanoparticles to the cutting fluid has been extensively tested for MQL applications, since the presence of nanoparticles usually enhances the base fluid lubrication and cooling capacities, thereby improving MQL technique performance [6, 7].

Zhang, D. et al. [8] compared the grinding of hardened AISI 1045 steel with MoS_2-based nanofluid applied via the MQL technique with the traditional MQL technique (oil without nanoparticles). The oil tested by the authors was a vegetable one (colza oil). The results showed that the MQL nanofluid at 1% concentration reduced the tangential grinding force, coefficient of friction, specific energy, and surface roughness (Ra parameter) by 27%, 21%, 28%, and 6%, respectively. According to the authors, the presence of nanoparticles in the cutting fluid improved the lubrication property and consequently the efficiency of the traditional MQL technique, which is attributed to the nanoparticle's anti-wear and anti-friction capability. The authors also tested different nanoparticle concentrations and observed a decrease in grinding performance (e.g., higher coefficient of friction, specific energy, and surface roughness) for volume concentrations over 2%. According to them, nanoparticles tend to cluster at high concentrations, which reduces lubrication performance.

De Paiva et al. [9] during the grinding of SAE 52100 hardened steel also evaluated the influence of graphene concentration dispersed in a semi-synthetic fluid applied with the MQL technique and found that, in terms of surface finish, graphene-based nanofluid was able to reduce surface roughness (Ra parameter) of ground surface up to 12% in comparison to the traditional MQL technique. Furthermore, they reported that the surface roughness of the ground surface increased with graphene concentration. According to the authors, high nanoparticle concentration diminishes the lubrication capacity of the cutting fluid due to particle agglomeration. Additionally, the agglomerated nanoparticles can act as abrasives at the contact zone, which tends to deteriorate the surface roughness of the ground surface.

Considering the importance of the type of fluid, type of nanoparticles, and their concentration to grinding performance, this work sought to evaluate the influence of base fluid and nanoparticles concentration on surface integrity of SAE 52100 hardened steel after grinding with graphene-based nanofluids applied with MQL technique. Two different and commercially available cutting fluids were used as the base fluid for graphene dispersion at different concentrations. The surface integrity of the workpiece after grinding was analyzed in terms of surface finish (Rq parameter) and microhardness below the machined surface.

2 Methodology

Grinding tests were performed on a MELLO P-36 peripheral surface grinding machine with a motor nominal power of 2.25 kW and a nominal speed of 2400 rpm. The resolution of the main spindle, the vertical axis in which the grinding wheel rotates, is 5 μm. A conventional white aluminum oxide (Al_2O_3) grinding wheel, from Norton-Saint Gobain Abrasives, has a 250 mm external diameter. Its designation is 38A60K6V. The workpiece material was the SAE 52100 hardened steel (60 ± 2 HRC) having a cylindrical geometry: 16 mm diameter and 17.4 mm height. Two parallel directly opposite surfaces of the workpiece were those used for the grinding tests. Worth mentioning that this particular material is widely employed for bearing applications and, therefore, is commonly submitted to the grinding process after heat treatment [10].

The cutting parameters employed for all grinding trials are grinding wheel speed (v_s) of 31 m/s, workspeed (v_w) of 7 m/min, and radial depth of cut (a_e) of 30 μm. Two different cutting fluids were tested in this work: a semi-synthetic vegetable-based oil (Vasco 7000) and a synthetic oil (Grindex 10), both from Blaser Swisslube. The fluids were used as base fluid to add multilayer graphene at two different concentrations: 0.025 wt.% and 0.075 wt.%. The multilayer graphene nanoparticles were produced by natural graphite exfoliation according to the methodology employed by Machuno et al. [11]. They consist of nanosheets with 1–30 nm thickness and 1–20 μm length. The graphene dispersion in the base fluids was performed using a commercially ultrasonic bath for two hours and a frequency of 40 kHz.

Six (6) different cutting fluids were tested in this work, all of them delivered via the MQL technique at a flow rate of 150 mL/h and 0.3 MPa of compressed air pressure. A conventional geometry nozzle was used, positioned to deliver the cutting fluid tangentially to the workpiece ground surface. The main characteristics of each base fluid as provided by the manufacturer shown in Table 1. For characterization, the thermal conductivity and kinematic viscosity of each cutting fluid tested were measured by using a Linseis THB-1 and an Anton Paar SVM 3000 viscosimeter, respectively. The results are shown in Table 2.

The grinding wheel was dressed before each grinding test by using a single-point diamond dresser, with 0.31 mm dressing width (b_d) and a dressing speed (v_d) of

Table 1. Main characteristics as provided by Blaser Swisslube for each base fluid.

	Semi-synthetic (Vasco 7000)	Synthetic (Grindex 10)
Color	Light brown	Yellow
Water content	-	40%
Esther content	25%	-
Density at 20 °C (g/cm^3)	0.98	1.12

Table 2. Thermal conductivity and kinematic viscosity at 40 °C for each cutting fluid tested in this work.

Cutting fluid	Graphene concentration (wt.%)	Thermal conductivity (Wm^{-1}K^{-1})	Kinematic viscosity (mm^{2}s^{-1})
Semi-synthetic	0 (without nanoparticles)	0.3137	71.8045
	0.025	0.2942	82.4005
	0.075	0.2939	81.1800
Synthetic	0 (without nanoparticles)	0.3548	10.6905
	0.025	0.3559	13.0490
	0.075	0.3615	13.9235

150 mm/min, which guarantee a grinding wheel overlap ratio (Ud) equal to 5. Each grinding test consisted in removing the total radial depth of cut of the plane surface of the workpiece through one passage of the grinding wheel, with one spark-out passage (with no depth of cut increment). Each grinding experiment was replicated once to increase the reliability of the results.

The surface integrity of the workpiece after grinding was evaluated in terms of surface finish (roughness Rq) and microhardness below the machined surface. A Taylor Hobson Surtronic S128 portable surface tester was employed for roughness measurements. Based on ISO 4288 [12], a 5.0 mm evaluation length, 0.8 sampling length, and Gaussian filter were used. Five surface roughness measurements were done for each grinding test and the mean and standard deviation values for the Rq parameter were selected for analysis. For microhardness measurements, a SHIMADZU HMV-2 microhardness tester was employed with Vickers indenter and parameters of 980.7 mN of load (HV0.1) for 10 s. The microhardness measurements were performed at 20 μm below the machined surface after standard metallographic preparation of samples (sanding using sandpapers with granulometry mesh #120, #200, #400, #600, and #1200, followed by polishing with 1.0 μm alumina paste). For each grinding trial, three measurements were taken, and the mean and standard deviation values were used for analysis.

3 Results and Discussions

The mean values of surface roughness (Rq parameter) for each cooling-lubrication condition tested in this work are shown in Fig. 1, where the error bars represent the standard deviation of the measurements. One notes that both base fluids (semi-synthetic and synthetic) presented the same results in terms of surface finish after grinding without nanoparticles. Considering the nanofluids, it can be noticed that for the lower concentration of nanoparticles (0.025 wt.%), the addition of graphene in the semi-synthetic fluid improved grinding performance by reducing Rq values by 8.6%. For the synthetic fluid, in contrast, a 60% increase in roughness Rq was observed. When machining at a higher concentration of nanoparticles (0.075 wt.%) dispersed in semi-synthetic and synthetic fluids, the presence of graphene led to an increase in Rq values of 7% and 91%, respectively.

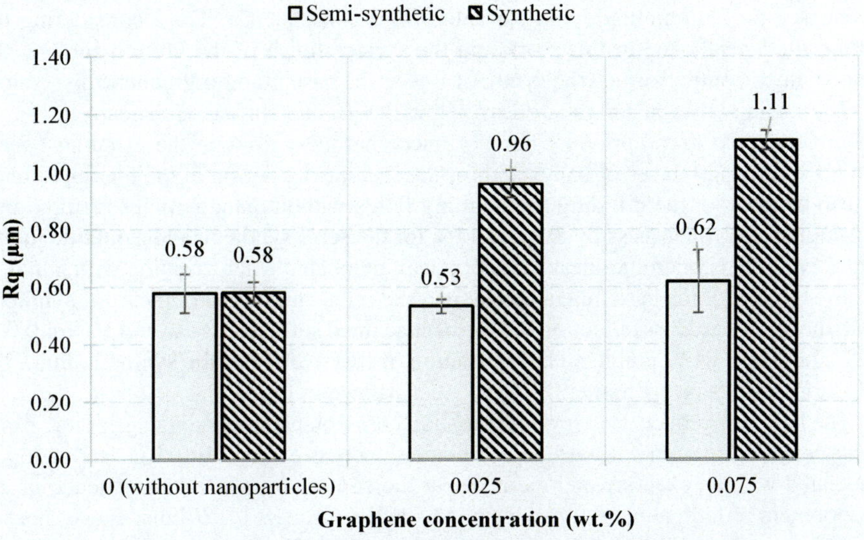

Fig. 1. Rq roughness of SAE 52100 hardened steel for the different cooling-lubrication conditions tested.

The benefits in terms of reducing surface roughness in grinding with nanofluids applied with the MQL technique are associated with the action mechanisms of nanoparticles at the contact zone, especially those which contributed to tribofilms formation. These protective films enhance tribological conditions between abrasive grains and the workpiece by reducing friction, which resulted in surface finishing improvement of components [13].

The action mechanisms of nanoparticles at the contact zone, however, are adversely affected in case of cluster formation in the nanofluid when a higher nanoparticles concentration is employed, which is associated with nanoparticle agglomeration that tends to occur due to high particle concentration, for instance. Zhang Y. et al. [14] evaluated different nanoparticle concentrations in MoS_2 nanofluids during the grinding of stainless steel with an Al_2O_3 wheel. The authors observed an increase in the coefficient of friction and the specific energy for nanoparticle concentrations over 6%. According to them, the dynamic stability of nanoparticles diminishes once their concentration in the base fluid increases, thereby reducing the lubrication capacity of the nanofluid. Thus, the increase in Rq values with graphene concentration, irrespective of base fluid, may be attributed to a possible nanoparticle's agglomeration, which is in good agreement with the findings from Zhang D. et al. [8], De Oliveira et al. [15] and De Paiva et al. [9].

Regarding the influence of the base fluid, it is worth mentioning that the synthetic fluid tested in this work presents water in its chemical composition as shown in Table 1. In this sense, the hyper hydrophobic characteristic of graphene [16, 17] may have favored agglomeration compared to the semi-synthetic fluid, which explains the worst results in terms of surface finish even for the lower nanoparticle concentration. Additionally, the abrasive-like behavior of nanoparticles is maximized with agglomeration [18, 19], which can lead to deeper and wider grooves in the ground surface, thereby increasing the surface profile amplitude and Rq values as a consequence. Thus, considering the cutting fluids evaluated in this work and the surface finish of the ground surface, the semi-synthetic outperformed the synthetic one as the base fluid for graphene dispersion.

The mean values of microhardness below the ground surface are shown in Fig. 2, where the dotted line represents the workpiece hardness prior to the grinding operation (801 HV), and the error bars stand for the standard deviation of the measurements. As can be noticed, the grinding with cutting fluids without nanoparticles resulted in a reduction in microhardness by 5% and 11% for the semi-synthetic and synthetic fluids, respectively. This microhardness reduction was attenuated when grinding with nanofluids, irrespective of the base fluid and nanoparticles concentration. For the semi-synthetic fluid, the microhardness reduction below the machined surface was 2% and 1% for 0.025 wt.% and 0.075 wt.% graphene concentration, respectively. For the synthetic fluid, the reductions were slightly higher: 5% and 3% as shown in Fig. 2.

The hardness reduction in ground parts is a common phenomenon in grinding, especially considering high-carbon hardened steels such as SAE 52100. This reduction is associated with an excessive tempering near the contact zone as a consequence of the development of high temperatures during the cutting process [1, 20] that causes heat to be more concentrated in the grinding zone and is mostly directed to the workpiece when a conventional abrasive grinding wheel is used, like that one employed in this work. An increase in hardness (rehardening) beneath the ground surface can also occur, which is a consequence of reaustenitization followed by the formation of untempered martensite [1]. Worth mentioning that the thermal damage extension beneath the ground surface, including the presence or not of untempered martensite, depends on the contact zone temperature level and its time of action. In terms of microhardness reduction (excessive tempering) for instance, an affected layer up to 180 μm below the machined surface was reported by De Paiva et al. [21] after grinding SAE 52100 hardened steel, with a 22% hardness reduction at 20 μm below the ground surface.

In this context, the microhardness results found in this work indicate that the presence of graphene in the cutting fluid contributed to better temperature control at the contact zone, thereby attenuating possible thermal damage to the workpiece in terms of hardness reduction beneath the machined surface. Worth mentioning that this better temperature control may be either associated with a reduction in temperature during grinding or a reduction in the heat partition that is conducted to the workpiece, which can be accomplished by friction reduction and/or heat dissipation enhancement. Although further analysis (e.g., cutting power and temperature measurements) must be carried out to better understand the graphene role in thermal damage reduction during grinding with nanofluids, the lamellar structure and outstanding thermal conductivity of graphene [22, 23] can provide better lubrication and cooling, respectively, which may explain the better temperature control at the contact zone.

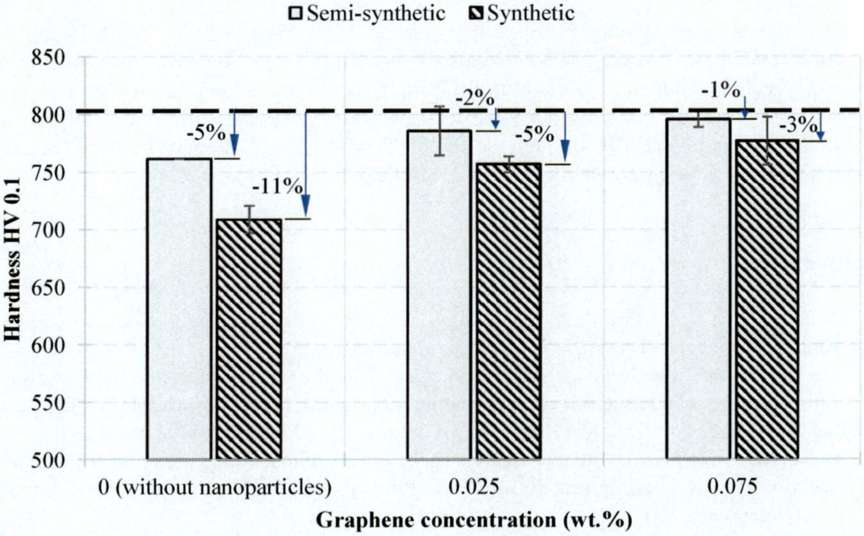

Fig. 2. Microhardness at 20 μm below the machined surface for the different cooling-lubrication conditions tested.

4 Conclusions

After grinding the SAE 52100 hardened steel with graphene nanofluids applied with the MQL technique using different base fluids and nanoparticles concentrations, the following conclusions can be drawn:

i. Both base fluid and nanoparticle concentration affected the surface integrity of the workpiece in terms of surface finish (Rq parameter) and microhardness below the machined surface.
ii. In terms of surface finish, the addition of graphene improved grinding performance by reducing Rq roughness (8.6% reduction) only for the semi-synthetic fluid at low graphene concentration (0.025 wt.%).

iii. Considering the cutting fluids evaluated in this work and the results of the surface finish of the ground surface, the semi-synthetic outperformed the synthetic one as the base fluid for graphene dispersion.
iv. Surface roughness of the ground surface increased with nanoparticle concentration irrespective of the base fluid.
v. For the conditions tested in this work, the presence of graphene in the cutting fluid contributed to attenuating thermal damage in terms of microhardness reduction below the machined surface. The best results were found after grinding with the higher nanoparticles concentration (0.075 wt.%), which promoted microhardness reductions of only 1% and 3% for semi-synthetic and synthetic fluids, respectively.

Acknowledgements. The authors would like to thank Blaser Swisslube for providing the cutting fluids tested in this work. The authors also thank Saint Gobain Abrasives for supplying the grinding wheel. The authors are grateful to CNPq, CAPES, and FAPEMIG for their financial support. Rosemar Batista da Silva acknowledges the CNPq (Processes No.: 311337/2016-3 and 426018/2018-4) and FAPEMIG through the process PPM-00492-18. Raphael Lima de Paiva also acknowledges the CNPq through the process Nº 140320/2016-4.

References

1. Malkin, S., Guo, C.: Grinding Technology - Theory and Applications of Machining With Abrasives. Industrial Press, New York, Second (2008)
2. Lima de Paiva, R., de SouzaRuzzi, R., Batista da Silva, R.: Contribution to the selection of cutting fluid type and its application technique for grinding of bearing steel. Proc. Inst. Mech. Eng. B J. Eng. Manuf., 1–11 (2021). https://doi.org/10.1177/09544054211040633
3. Hadad, M.: An experimental investigation of the effects of machining parameters on environmentally friendly grinding process. J. Clean. Prod. **108**, 217–231 (2015). https://doi.org/10.1016/j.jclepro.2015.05.092
4. Sanchez, J.A., Pombo, I., Alberdi, R., et al.: Machining evaluation of a hybrid MQL-CO2 grinding technology. J. Clean Prod. **18**, 1840–1849 (2010). https://doi.org/10.1016/j.jclepro.2010.07.002
5. Walker, T.: The MQL Handbook – A guide to machining with Minimum Quantity Lubrication. Unist Inc, Grand Rapids, MI (2013)
6. Devendiran, D.K., Amirtham, V.A.: A review on preparation, characterization, properties and applications of nanofluids. Renew. Sustain. Energy Rev. **60**, 21–40 (2016). https://doi.org/10.1016/j.rser.2016.01.055
7. Sharma, V.S., Singh, G., Sorby, K.: A review on minimum quantity lubrication for machining processes. Mater. Manuf. Processes **30**, 935–953 (2015). https://doi.org/10.1080/10426914.2014.994759
8. Zhang, D., Li, C., Jia, D., et al.: Specific grinding energy and surface roughness of nanoparticle jet minimum quantity lubrication in grinding. Chin. J. Aeronaut. **28**, 570–581 (2015). https://doi.org/10.1016/j.cja.2014.12.035
9. de Paiva, R.L., de Souza Ruzzi, R., de Oliveira, L.R., et al.: Experimental study of the influence of graphene platelets on the performance of grinding of SAE 52100 steel. Int. J. Adv. Manuf. Technol. **110** (2020). Doi:https://doi.org/10.1007/s00170-020-05866-x
10. Balart, M.J., Bouzina, A., Edwards, L., Fitzpatrick, M.E.: The onset of tensile residual stresses in grinding of hardened steels. Mater. Sci. Eng., A **367**, 132–142 (2004). https://doi.org/10.1016/j.msea.2003.10.239

11. Machuno, L.G.B., Oliveira, A.R., Furlan, R.H., et al.: Multilayer graphene films obtained by dip coating technique. Mater. Res. **18**, 775–780 (2015). https://doi.org/10.1590/1516-1439.005415
12. ISO 4288:1996 (1996) Geometrical Product Specifications (GPS) — Surface texture: Profile method — Rules and procedures for the assessment of surface texture. 8
13. Singh, H., Sharma, V.S., Singh, S., Dogra, M.: Nanofluids assisted environmental friendly lubricating strategies for the surface grinding of titanium alloy: Ti6Al4V-ELI. J. Manuf. Process. **39**, 241–249 (2019). https://doi.org/10.1016/j.jmapro.2019.02.004
14. Zhang, Y., Li, C., Jia, D., et al.: Experimental evaluation of MoS2 nanoparticles in jet MQL grinding with different types of vegetable oil as base oil. J. Clean. Prod. **87**, 930–940 (2015). https://doi.org/10.1016/j.jclepro.2014.09.027
15. De Oliveira, D., Da Silva, R.B., Gelamo, R.V.: Influence of multilayer graphene platelet concentration dispersed in semi-synthetic oil on the grinding performance of Inconel 718 alloy under various machining conditions. Wear **426–427**, 1371–1383 (2019). https://doi.org/10.1016/j.wear.2019.01.114
16. Nguyen, D.D., Tai, N.H., Lee, S.B., Kuo, W.S.: Superhydrophobic and superoleophilic properties of graphene-based sponges fabricated using a facile dip coating method. Energy Environ. Sci. **5**, 7908–7912 (2012). https://doi.org/10.1039/c2ee21848h
17. Alberts, M., Kalaitzidou, K., Melkote, S.: An investigation of graphite nanoplatelets as lubricant in grinding. Int. J. Mach. Tools Manuf. **49**, 966–970 (2009). https://doi.org/10.1016/j.ijmachtools.2009.06.005
18. Lee, C.-G., Hwang, Y.-J., Choi, Y.-M., et al.: A study on the tribological characteristics of graphite nano lubricants. Int. J. Precis. Eng. Manuf. **10**, 85–90 (2009). https://doi.org/10.1007/s12541-009-0013-4
19. Azman, S.S.N., Zulkifli, N.W.M., Masjuki, H., et al.: Study of tribological properties of lubricating oil blend added with graphene nanoplatelets. J. Mater. Res. **31**, 1932–1938 (2016). https://doi.org/10.1557/jmr.2016.24
20. Seidel, M.W., Zösch, A., Härtel, K.: Grinding burn inspection: tools for supervising and objectifying of the testing process. Forschung im Ingenieurwesen/Engineering Research **82**, 253–259 (2018). https://doi.org/10.1007/s10010-018-0270-4
21. de Paiva, R.L., de Souza, R.R., da Silva, R.B.: An approach to reduce thermal damages on grinding of bearing steel by controlling cutting fluid temperature. Metals (Basel) **11**, 1660 (2021). https://doi.org/10.3390/met11101660
22. Liu, J., Wang, F., Zhang, L., et al.: Thermodynamic properties and thermal stability of ionic liquid-based nanofluids containing graphene as advanced heat transfer fluids for medium-to-high-temperature applications. Renew Energy **63**, 519–523 (2014). https://doi.org/10.1016/j.renene.2013.10.002
23. Zhang, G., Häusler, I., Österle, W., et al.: Formation and function mechanisms of nanostructured tribofilms of epoxy-based hybrid nanocomposites. Wear **342–343**, 181–188 (2015). https://doi.org/10.1016/j.wear.2015.08.025

Influence of Cutting Speed on Machining Force and Chip Morphology in Turning Self-lubricating Composites Sinterized

Caroline Francisco Dorneles(✉), Sara Builes Díaz, Felipe Gustavo Ebersbach, and Rolf Bertrand Schroeter

Federal University of Santa Catarina, s/n Eng. Agronômico Andrei Cristian Ferreira st, Trindade, Florianópolis, SC 88040-900, Brazil
carol.fdorneles@gmail.com

Abstract. Currently, a constant concern in the industrial sectors is looking for an improvement in energy efficiency for machinery and equipment. In this context, due to the satisfactory mechanical properties and low coefficient of friction, sintered self-lubricating composites are a viable alternative to materials already used in engineering to contribute to a decrease in energy consumption and increase in sustainability. Hence, this work presents an analysis of the effect of varying cutting speed on the components of machining force and chip morphology in external longitudinal turning of a sintered self-lubricating composite. The specimens were manufactured by powder metallurgy process and have a metallic matrix, composed of iron powder with the addition of alloying elements (Ni and Si). hBN and graphite particles are used as solid lubricants. Six cutting speeds (v_c) were adopted, from 350 m/min to 850 m/min, with an increment of 100 m/min. The depth of cut and feed were set at 1.0 mm and 0.2 mm, respectively. The turning tests were carried out using carbide inserts, class P20, triangular shape with three cutting edges on a single face. During the machining tests, machining force were measured, and chip samples were collected. As a result, it was possible to observe a decrease in the components of the machining force with the increase of the cutting speed. Furthermore, when analyzing the chip morphology, the formation of segmented/arc chips was observed. As the cutting speed increases, the chips tend to present a reduction in their arc length, which modify the shape.

Keywords: Sintered self-lubricating composite · Machining force · Chip morphology

1 Introduction

Within the current context of sustainability, energy efficiency has become one of the most important research topics. Recent work indicates that approximately 30% of energy resources around the world are consumed to overcome friction between mechanical components [1]. Losses due to this phenomenon represented between 2 and 7% of GDP worldwide, just in 2018. Holmberg and Erdemir [2] state that approximately 23% of the

D. de Oliveira et al. (Eds.): COBEF 2023, LNME, pp. 210–217, 2024.
https://doi.org/10.1007/978-3-031-43555-3_20

energy produced in the world is applied to supply defective tribological contacts, with 20% used to overcome friction and 3% to recover worn parts. Therefore, research with the aim of improving the tribological performance of surfaces, especially those that act to reduce the coefficient of friction, are obtaining more and more importance in this context.

According to Gwidon and Stachowiak [3], sintered self-lubricating composites are a good solution for high pressure and temperature applications, where liquid lubrication does not perform well. These materials are fabricated by powder metallurgy process and are characterized by two chemically different phases, which are a ferrous metallic matrix, with the addition of alloying elements, and particles of solid lubricants dispersed in its volume. Edemir [4] explains that solid lubricants are materials with low coefficient of friction and low molecular compression, capable of creating layers due to the ease of being sheared, Therefore, they prevent contact between surfaces in relative motion.

With the aim of improving the surface quality of self-lubricating composites, the machining process is used as a finishing process after the material's sintering stage. Within this context, the turning process becomes an excellent alternative due to its wide application in industry.

The discontinuity of the self-lubricating composite matrix in combination with their mechanical properties, have an influence when machining, especially in the quality of the machined surface. Sintered materials have low thermal conductivity, due to the high porosity (inherent to the process), which results in an increase in temperature in the cutting zone during machining [5]. According to M'saoubi et al. [6], Nieslony and Kiszka [7] and Grzesik [8], the porosity present in sintered materials causes a decrease in forces during machining, which also influences chip formation during the process.

Ebersbach et al. [9] evaluated the effect of changing the cutting speed on the behavior of the machining force components and on the morphology of the chips generated in the turning of sintered self-lubricating composites. The tests were performed varying the cutting speed (v_c = 100 and 200 m/min), advance (f = 0.1 and 0.2 mm), cutting depth (a_p = 0.5 and 1 mm), as well as the manufacturing route of single compaction (SP) and double compaction (DP) materials. The authors observed that regardless of the material manufacturing route, for both depth of cut and feed tested, the increase in cutting speed resulted in a reduction in cutting force values of about 10%. The same behavior was observed for passive force, showing a reduction of approximately 25%. In addition, the type of chip obtained was segmented, and the chip thickness ratio was influenced by the variation in cutting parameters and the manufacturing route of the materials.

In this context, this work presents research on the effect of cutting speed on the components of machining force and chip morphology during external longitudinal turning of a sintered self-lubricating composite.

2 Materials and Methods

For the production of composites by powder metallurgy, the materials that were used were pure iron powder as raw material for the ferrous matrix (Höganäs, AHC 100.29, d50 = 100 μm). For increasing matrix strength property was added alloying elements, such as carbon and nickel. Hexagonal boron nitride (Momentive, AC6028, d50 = 125 μm) was added to the alloy along with graphite (Nacional Grafite, Micrograph 99545HP, d50 = 32 μm) to promote dry lubrication. However, it is necessary to use silicon (Osprey Sandvik, alloy Fe45Si, d50 = 10 μm) as an alloying element for the stabilization of alpha iron and hardening of the matrix. In Table 1 is described the alloy chemical composition. Using balance in iron, was used 0.8% by mass of carbon and 0.5% by mass of Fe4Si together with 1% and 6.5% by volume of hexagonal boron nitride (d50 = 125 μm) and graphite (d50 = 32 μm), respectively.

Table 1. Powders used to manufacture the sintered self-lubricating composite.

Powder	Commercial name	Fabricant	Size [μm]
Iron	AHC 100,29	Höganäs	100
hBN	hBN-6028	Momentive	100–150
Grafit	Micrograf 99545 HP	Nacional Grafite	20–45
Nickel	INCO 123	INCO	22
Fe45Si	Alloy Fe45Si	Sandvik	10

To mix the powders a Y-type mixer was used during 45 min at 35 rpm. After the mixing, the powders were pressed in a double action hydraulic press. The materials were produced by simple compaction (SC using a pressure of 600 MPa). The sintering of the materials was carried out using a tubular furnace in an atmosphere of 95% argon and 5% hydrogen with 2 temperature stages. In the first stage, a heating rate of 5 °C/min and a stay of 30 min in a temperature of 550 °C were used, In the second stage, the final sintering was isothermal, at a temperature of 1125 °C for 60 min, using a heating rate of 10 °C/min. The average hardness of the materials was 123 HB. The maximum tensile stress and yield stress were 205 MPa and 197 MPa, respectively.

After the sintering process, the specimens were drilled and reamed in order to produce a longitudinal through-hole with keyway fitting. It was made for subsequent assembly of the specimen in the machine tool fastening system. The specimen had disk form, with a diameter of 65 mm and a width of 20 mm, with a 16 mm through hole (quality H7) and a keyway.

The material microstructure was characterized by metallographic analysis, an optical microscope Leica model DM4000 M/DFC45 was used for this. In Fig. 1, it is possible to observe the different phases present in the composite, such as ferrite, around the pores, and pearlite; in addition to phases rich in nickel and cementite, which make the material heterogeneous and discontinuous.

Fig. 1. Self-lubricant sintered composite microstructure.

2.1 Turning Process

For turning the specimens, an external longitudinal turning process was carried out. The machine in use was a Heiligenstaedt CNC lathe, Heynumat 10 model, with spindle drive power of 70 kW and maximum rotation of 4500 rpm. The equipment has a hydraulic closure of three jaws with a pressure of 45 bar and Siemens 802D CNC control. The materials were machined with carbide inserts of P20 class, with three cutting edge in a triangular geometry, code TCMT 16T308-SM IC907, without coating, and with a cutting edge radius (ρ) of 47 μm. The inserts were supplied by the company ISCAR do Brasil, as well as the tool holder used (code STGCL 2020K-16). The insert had a nose radius of 0.8 mm, a rake angle of 18° and a relief angle was 7°. The insert was fixed with a side rake angle and approach angle of 0° and 91°, respectively. The tests were carried out dry, varying the cutting speed (v_c) in 6 levels: 350, 450, 550, 650, 750 and 850 m/min. The depth of cut (a_p) and feed (f) was 1.0 mm and 0.2 mm, respectively. Two replicates of each test were performed.

The measurement of the machining force components (cutting force, feed force and passive force) was performed during all tests using a piezoelectric platform, manufactured by the company Kistler Instrument AG, model 9257 A, with the aid of amplifiers, also from a Kistler, models 5011 and 5006, connected to a National Instruments acquisition board, model NI USB 6218. The values of the machining force components were obtained by the difference between the average of the signals obtained during cutting and the average of the signals obtained in empty, adopting the measurement of force with relative values. Furthermore, an acquisition rate of 1 kHz was used for the three components of the machining force.

For tests with cutting speeds of 350, 650 and 850 m/min, samples of the chips generated during machining were collected. The morphological analysis of the chip was performed using images obtained by a Scanning Electron Microscope (SEM), Tescan brand, model VEGA3. Chip length was measured using ImageJ software.

3 Results and Discussion

3.1 Machining Force

The mean values of the machining force components with their respective confidence intervals are shown in Fig. 2. It is possible to observe that the increase in the cutting speed provided a reduction in the values of the machining force components, for the tested cutting speeds. Increasing the cutting speed from 350 to 850 m/min led to a decrease of 7% in the feed force, 5% in the passive force and 9% in the cutting force. However, at the cutting speed of 650 m/min, there was an increase approximately of 3% for passive force and 10% for cutting force in relation to the speed of 350 m/min. The variation in cutting speed did not show significant influence on feed force and passive force with any of the cutting speeds (P-value equal to 0.46 and 0.25, respectively). For the cutting force, there was a significant difference between the cutting speeds of 350 and 850 m/min (p-value = 0.001), reinforcing the behavior observed.

As explained by Denkena et al. [10], the increase in passive and forward forces, for a cutting speed of 650 m/min, can be explained by the fact that this region presents a transient behavior due to phenomena in the cutting zone, such as temperature increase, rapid flow out of the chip and increased vibrations in the process.

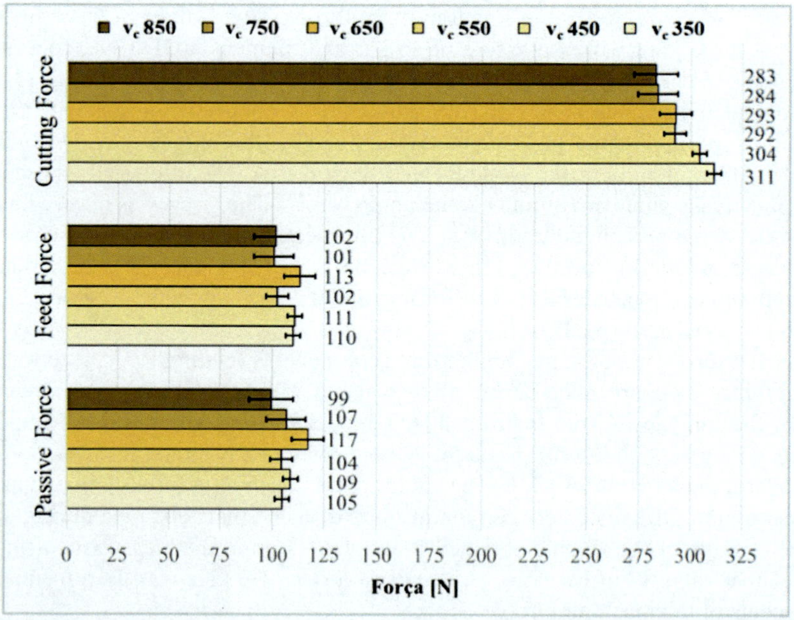

Fig. 2. Mean values of the components of the machine force with the cutting speed variation.

According to Klocke [11], cutting speed is a process parameter that directly influences the temperature of the cutting region. The increase in cutting speed leads to an increase in temperature in the cutting zone of the material, and consequently, there is a loss of mechanical strength. This makes it easier to cut the material, thus reducing the machining force components.

Two types of phenomena can also contribute to the decrease in cutting force with increasing cutting speed. According to Wang et al. [12], with an increase in cutting speed, there is an increase in temperature in the cutting zone, which, in addition to decreasing the mechanical strength of the material, can provide welding of the material cut at the chip-tool interface. This fact is capable of generating a film that reduces the coefficient of friction due to the greater ease of flow through the face of the tool, therefore, it could be providing a greater reduction of the cutting force. However, there is a difference when even higher cutting speeds are used, as the melting phenomenon is not dominant due to the very high deformation and embrittlement of the chip, in addition to the rapid exit from the cutting zone. Consequently, there is a decrease in adhesion of the workpiece material at the chip-tool interface. This phenomenon also helps to reduce the coefficient of friction, hence producing a reduction in the forces acting in the process.

3.2 Chip Morphology

Figure 3 show the chip morphology for cutting speed conditions (v_c) of 350, 650 and 850 m/min, respectively. Note that with increasing cutting speed there is a tendency to reduce chip length. The dotted line in Fig. 3. Show the measured perimeters. A mean value of 3.6 $\pm$ 1.7 mm was obtained for the cutting speed of 350 m/min and of 2.7 $\pm$ 0.9 mm for a cutting speed of 850 m/min, it was found that there is no significant difference between the lengths (P value = 0.212). Furthermore, it is possible to observe that for the cutting speed of 850 m/min the chips presented less curvature.

According to analyzes carried out by Ebersbach et al. [9], for the sintered self-lubricating composite, the sheared chip is characteristic of the cut due to the heterogeneity and discontinuity of the material. The difference in the mechanical properties of the different phases facilitates fracture by the phase of least mechanical property. Therefore, despite presenting this behavior even at the lowest cutting speed tested, Molnar et al. [13] explain that the trend towards increased fracture frequency is also a cause of increased data dispersion in the measurement of machining force components.

One of the consequences of increasing cutting speed is the change in chip type. According to He et al. [14], the high strain rates generated by the removal speed cause the frequency of the chip shear bands to increase and the material to adopt a more brittle behavior due to strain hardening. Therefore, when cutting the sintered self-lubricating composite, the high strain rates generated by the increase in cutting speed could be one of the causes of the decrease in chip length in addition to the change to a format with less curvature. Other reasons that may be influencing the increase in chip fracture could be related to the loss of stability and increased vibrations of the machine-tool system when the cutting speed is increased.

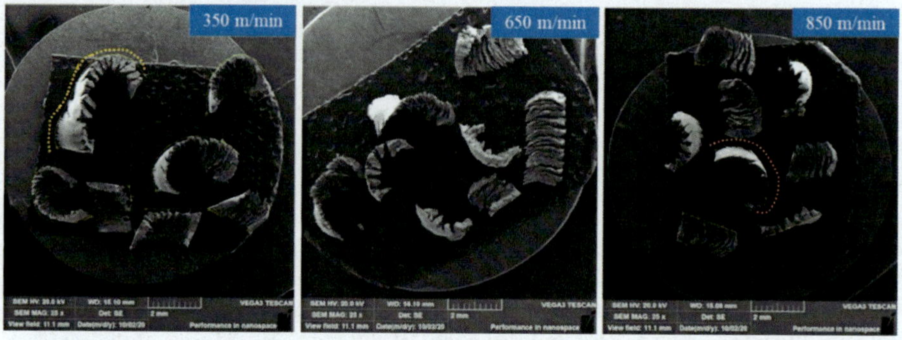

Fig. 3. Chip morphology using cutting speed of 350, 650 and 850 m/min.

4 Conclusion

According to the results presented during this work, it was possible to observe the effects of increasing the cutting speed on the behavior of the machining force components and on the chip morphology of sintered self-lubricating composite materials.

The variation in the cutting speed directly influenced the cutting force, when the cutting speed was increased from 350 to 850 m/min there was a reduction of 7%. For the components of passive force and feed force, no significant influence was presented, however, it was observed a tendency of decrease of the values of the components with the increase of the cutting speed, and it was represented by a percentage of approximately 9% and 5% respectively. With the increase in cutting speed, there is an increase in temperature in the cutting zone, with a reduction in the mechanical properties of the material and thus, greater ease for the cutting edge of the tool to penetrate the material and perform the cut.

Regarding chip morphology, despite showing a decrease in curvature and chip length, no drastic change was observed, nor a statistically significant difference found. The phenomena involved in the trend of decreasing chip length could be related to the increase in vibrations in the system due to the increase in cutting speed.

Acknowledgment. The authors thank the Federal University of Santa Catarina (UFSC), especially the Graduate Program in Mechanical Engineering (POSMEC); to the Precision Mechanics Laboratory (LMP), at UFSC; to the Materials Laboratory (LABMAT), at UFSC. The authors would also like to thank the National Council for Scientific and Technological Development (CNPq), the Coordination for the Improvement of Higher Education Personnel (CAPES) and FUMDES/UNIEDU, as well as Embraco and Iscar do Brasil for funding this research.

References

1. Luo, J., Zhou, X.: Future industry nearly getting rid of wear and frictional energy consumption. Superlubricitive Eng. Friction **8**, 643–665 (2020)
2. Holmberg, K., Erdemir, A.: The impact of tribology on energy use and CO2 emission globally and in combustion engine and electric cars. Tribol. Int. **135**, 389–396 (2019)

3. Gwidon, W., Stachowiak, A.W.B.: Engineering Tribology. Elsevier. United Kingdom (2006)
4. Erdemir, A.: Solid lubricants and self-lubricating films. Modern Tribology Handbook: Principles of Tribology **1**, 787–825 (2000)
5. Šalak, A., Vasiko, K., Selecka, M., Danninger, H.: New short time face turning method for testing the machinability of PM steels. J. Mater. Process. Technol. **176**, 62–29 (2006)
6. M'saoubi, R., Czotscher, T., Andersson, O., Meyerb, D.: Machinability of powder metallurgy steels using PcBN inserts. Procedia CIRP **14**, 83–86 (2014)
7. Nieslony, P., Kiszka, P.: An investigation of surface texture after turning PM armco iron. Procedia CIRP (1), 671 (2012)
8. Grzesik, W.: Prediction of the functional performance of machined components based on surface topography: state of the art. J. Mater. Eng. Perform. **25**, 4460–4468 (2016)
9. Ebersbach, F.G., et al.: Effect of cutting parameters in machining force, surface texture and chips morphology obtained in turning of sintered self-lubricating composites. Mater. Res. **23**, 1–10 (2020)
10. Denkena, B., Ben Amor, R., De León-García, L., Dege, J.: Material specific definition of the high speed cutting range. Int. J. Mach. Machinability Mater. (2), 176–185 (2007)
11. Klocke, F.: Manufacturing Processes 1. Springer, Heidelberg (2011)
12. Wang, B., et al.: Advancements in material removal mechanism and surface integrity of high-speed metal cutting: A review. Int. J. Mach. Tools Manufacture (166), 103744 (2020)
13. Molnar, T.G., Berezvai, S., Kiss, A.K, Bachrathy, D., Stepan, G.: Experimental investigation of dynamic chip formation in orthogonal cutting. Int. J. Mach. Tools Manuf. (145), 103429 (2019)
14. He, Y., Zhang, J., Qi, Y., Liu, H., Memon, A.R., Zhao, W.: Numerical study of microstructural effects on chip formation in high speed cutting of ductile iron with discrete element method. J. Mater. Process. Technol. **249**, 291–301 (2017)

Numerical Optimization of Cutting Parameters for Texturization of Dental Implants

Tarcísio Ladeia de Oliveira(✉), Renato Goulart Jasinevicius, Jonas de Carvalho, and Alessandro Roger Rodrigues

Sao Carlos School of Engineering, University of Sao Paulo (EESC-USP), Av. Trabalhador São-Carlense, 400, Parque Arnold Schimidt, São Carlos, SP 13566-590, Brazil
tarcisio.oliveira@usp.br

Abstract. The surface texture of implants was shown to be an important factor in decreasing the intensity of the body's immunological response towards them and guaranteeing their longevity. Currently, regarding dental implants, an average surface roughness (S_a) between 1 and 2 μm is the norm, as that allows for better osseointegration. Many kinds of surface treatments may be used to obtain roughness in that range, each one having its own characteristics and roughness ranges. However, it should also be possible to obtain such surfaces using ultrasonic elliptical turning, but in order to obtain the desired surface, it is necessary to obtain the right set of cutting parameters. This paper, therefore, aims to develop a method to calculate the cutting parameters needed to such surface, basing itself on topography simulation and conventional optimization. The method consists on using shape functions which are capable of reproducing the shape of the surface to be obtained in function of the cutting parameters (being based specifically on feed rate and cutting speed) and using gradient-based optimization methods to find the optimal parameters. The method was implemented and tested considering a problem of maximization of surface area with a roughness constraint, problem developed considering the needs of dental implants. The implementation was shown to be capable of consistently obtaining a surface with the desired characteristics. However, a simplified mathematical model was used, and experimental validation was not done, matters which shall be dealt with in future research.

Keywords: Optimization · Cutting Parameters · Dental Implants · Turning

1 Introduction

One of the crucial steps necessary for long-term retention of a dental implant is the development of secondary stability, also known as *osseointegration* [1]. Secondary stability consists on the development of biomechanical stability as a result of the regeneration of bone around the implant, and, consequently, in rigid retention of the implant. It is complementary to primary stability, which is developed from the implantation methodology and the implant's macrostructure [1].

D. de Oliveira et al. (Eds.): COBEF 2023, LNME, pp. 218–230, 2024.
https://doi.org/10.1007/978-3-031-43555-3_21

After implantation, the bone begins to remodel itself around the implant. The trauma from implantation causes progressive tissue necrosis due to tissue compression, decreasing stability, and the regeneration process begins at about the same time [2]. Consequently, this adjustment process results in a temporary decrease in overall stability (Fig. 1), but it may lead to implant rejection if it is sufficiently protracted or acute.

Therefore, minimizing the intensity of such stability drop is necessary to avoid early implant failure. This may be done by combining two approaches: maximizing primary stability and accelerating the development of osseointegration. The first approach is commonly based on using a specialized macrostructure (such as screw features and implant shape) [3] as well as improving implantation methodology [1]. The second approach involves developing a more favorable environment for faster bone regeneration.

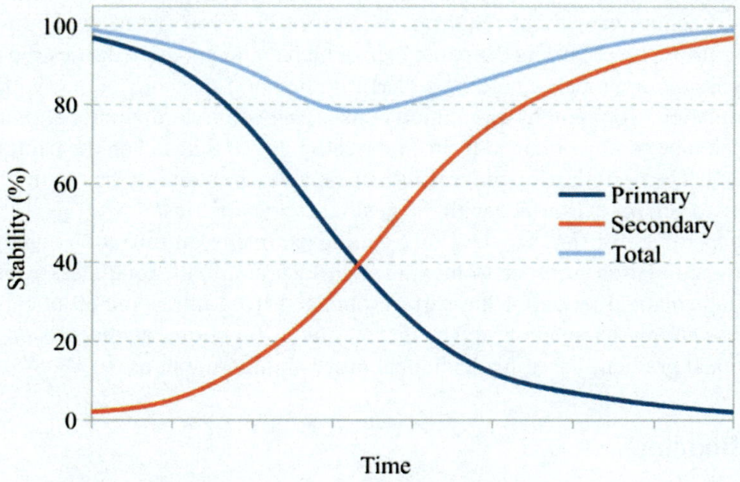

Fig. 1. Illustration of the evolution of implant stability over time.

One of the approaches used to prepare such environment for osseointegration consists on texturing the implant surface [4]. The surface features affect factors such as wettability, which in turn affects the adhesion of proteins and tissue on the surface [5], and surface area, which increases the contact area between bone and implant and makes the attachment more rigid [3]. Surface roughness (S_a) affects both factors, therefore, it is one of the main fields of research. Lower roughness tends to result in lower surface area and lower wettability. Consequently, using greater roughness would be ideal.

On the other hand, increasing roughness might also make it easier for bacteria to adhere and form a bacterial biofilm, which is hard to remove, and may consequently result in early implant failure and antibiotic-resistant infections [6]. Due to that, dental implants currently tend to have S_a between 1 and 2 μm [1, 7].

In order to obtain roughness within that range, some kinds of surface treatments tend to be used, such as grit-blasting, acid-etching, and anodization, as, in implantology, fabrication by turning is commonly associated with smooth ($S_a < 0.5$ μm) and minimally rough (0.5 μm $< S_a <$ 1 μm) surfaces [8]. However, the turning process has a much wider surface roughness range, especially vibration-assisted turning, which is commonly used to assist in the cutting of difficult-to-machine materials, such as commercially pure titanium [9].

Vibration-assisted turning consists on applying vibration systems onto the turning process (commonly onto the tool) as a way to assist in metal cutting. As a consequence of that vibratory motion, tool life increases, the roughness range is wider (it is possible to obtain average roughness values as low as 10 nm) and the dimensional accuracy is increased [9]. Ultrasonic elliptical turning is one kind that should be highlighted, as it is already being applied on implant fabrication and results in considerable improvements in machinability and surface finish [10].

Although it is commonly used to obtain smooth surfaces, ultrasonic elliptical turning may also be used to develop textures with arbitrary roughness by selecting the appropriate cutting and vibration parameters [11]. This way, the dental implants could be fabricated purely by machining, without requiring post-processing for texturing. The problem, however, consists on obtaining the parameters which could generate the desired surface.

One possible approach would be to use optimization methods, along with topography simulation. Topography simulation consists on using computational methods to simulate the topography obtained from a fabrication process based on the parameters to be used [12]. The optimization process is then capable of obtaining the required cutting parameters to generate a surface with the desired characteristics.

This paper therefore aims to develop a cutting parameter optimization method using topography simulation in order to maximize surface area with roughness constraints, using a mathematical model of the surface obtained from ultrasonic elliptical turning based on the cutting parameters, and using continuous functions, so that it is possible to use traditional gradient-based mathematical programming methods.

2 Methodology

The research being done is currently in its early stages, being purely theoretical and computational in its current state, but it will be subject to experimental validation in the near future. The mathematical modeling of the resulting geometry is based on a simplified model of ultrasonic elliptical turning, but it will be improved upon in order to represent future experimental results with greater accuracy.

2.1 Mathematical Model

The mathematical model is based on the sum of three features of the turning process, which were modeled as independent: material removal along the x axis with the shape of the cutting tool, forming a rectilinear pattern which is repeated along the y axis, the generation of the ellipsoidal texture along the x axis, and the modeling of random vibrations along the x axis. The space between peaks is defined by the feed rate f, and the spacing between ellipses, by the distance Δ_{uet}, related to the cutting velocity v_c. Figure 2 illustrates this model.

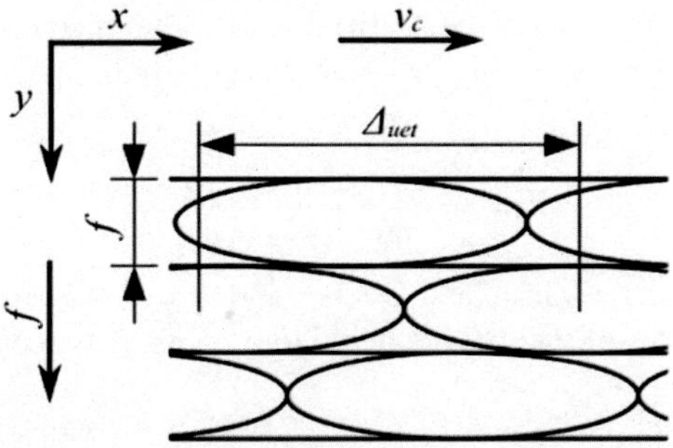

Fig. 2. Illustration of the mathematical model of the machined surface. The parallel lines represent the peaks generated by the sides of the tool, and the ellipses, the boundaries of the ellipsoidal sections generated from elliptical vibration.

Tool Model. The tool was modeled as a triangle with a rounded tip of radius r and angles α_1 and α_2 (Fig. 3).

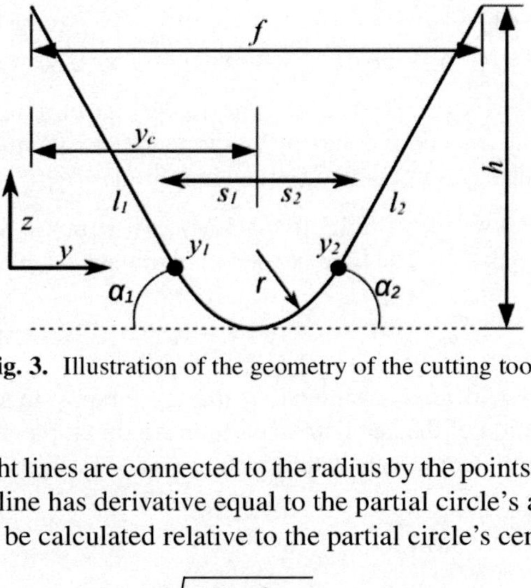

Fig. 3. Illustration of the geometry of the cutting tool.

The lateral straight lines are connected to the radius by the points (y_1, z_1) and (y_2, z_2). Assuming that each line has derivative equal to the partial circle's at the point that they meet, y_1 and y_2 may be calculated relative to the partial circle's center by:

$$y_i = \sqrt{\frac{r^2 \tan^2 \alpha_i}{\tan^2 \alpha_i + 1}}, i = 1, 2. \tag{1}$$

From that, it is possible to define the lines $l_1(y) = -\tan \alpha_1 y + b_1$ and $l_2(y) = \tan \alpha_2 y + b_2$ from the curve with radius r (that is, $c(y) = -\sqrt{r^2 - y^2} + r$) by:

$$\begin{aligned} b_1 &= -\sqrt{r^2 - y_1^2} + r + \tan \alpha_1 y_1 \\ b_2 &= -\sqrt{r^2 - y_2^2} + r - \tan \alpha_2 y_2 \end{aligned} \tag{2}$$

To correctly position the geometry, it is necessary to define a distance to the center y_c so that both sides of the geometry have the same height h. To do so, it is first necessary to know if f is: (a) greater than $y_2 - y_1$; (b) less than that distance; or, in case $\alpha_1 \neq \alpha_2$, (c) $-y_c > y_1$ and $f - y_c < y_2$ or $-y_c < y_1$ and $f - y_c > y_2$$ First, y_c is calculated assuming case (a):

$$y_c = \frac{\tan\alpha_2 f + b_2 - b_1}{\tan\alpha_1 + \tan\alpha_2}. \tag{3}$$

After y_c is obtained, the variables $s_1 = -y_c$ and $s_2 = f - y_c$ are defined, representing the lengths of both sides of the tool, centered on y_c, and y_c is recalculated by:

$$\begin{cases} y_c = \frac{\tan\alpha_2 f + b_2 - b_1}{\tan\alpha_1 + \tan\alpha_2}, & \text{if } s_1 < y_1 \text{ and } s_2 > y_2 \\ y_c = \frac{f}{2}, & \text{if } s_1 > y_1 \text{ and } s_2 < y_2 \\ \tan\alpha_1 y_c + b_1 = -\sqrt{r^2 - (f - y_c)^2} + r, & \text{if } s_1 < y_1 \text{ and } s_2 < y_2 (\text{case } \alpha_1 < \alpha_2) \\ \tan\alpha_2 (f - y_c) + b_2 = -\sqrt{r^2 - y_c^2} + r, & \text{if } s_1 > y_1 \text{ and } s_2 > y_2 (\text{case } \alpha_1 > \alpha_2) \end{cases} \tag{4}$$

After y_c is calculated, it is possible to obtain the height h by:

$$h = \begin{cases} \tan\alpha_1 y_c + b_1, & \text{if } s_1 < y_1 \text{ and } s_2 > y_2 \\ -\sqrt{r^2 - y_c^2} + r, & \text{if } s_1 > y_1 \text{ and } s_2 < y_2 \\ \tan\alpha_1 y_c + b_1, & \text{if } s_1 < y_1 \text{ and } s_2 < y_2 (\text{case } \alpha_1 < \alpha_2) \\ \tan\alpha_2 (f - y_c) + b_2, & \text{if } s_1 > y_1 \text{ and } s_2 > y_2 (\text{case } \alpha_1 > \alpha_2) \end{cases} \tag{5}$$

Elliptical Movement Model. The texture generated by the elliptical movement is modeled by generating ellipse sections of height z_{uet} along the x axis, with length Δ_{uet} (Fig. 4). The elliptical movement has frequency f_{uet} and amplitudes A_{Zuet} and A_{Xuet}.

From Brehl and Dow [9], with the frequency f_{uet} it is possible to define the tool's rotation speed as $\omega = 2\pi f_{uet}$. The time needed to complete a cycle is, therefore:

$$\omega T = 2\pi \Rightarrow T = \frac{1}{f_{uet}}. \tag{6}$$

That, however, would require simulating the topography in function of time. To accelerate the generation of the geometry, and to make the implementation more parallelizable, it is more efficient to calculate it in function of position. We can then define the time t as:

$$t = \frac{x}{v_c}, \tag{7}$$

consequently, the distance that the tool travels in period T (that is, Δ_{uet}) can be obtained by:

$$\Delta_{uet} = \frac{v_c}{f_{uet}}, \tag{8}$$

and the height z_{uet} can be obtained from the ellipse equation:

$$z_{uet} = A_{Zuet}\left[1 - \sqrt{1 - \left(\frac{\Delta_{uet}}{2A_{Xuet}}\right)^2}\right]. \tag{9}$$

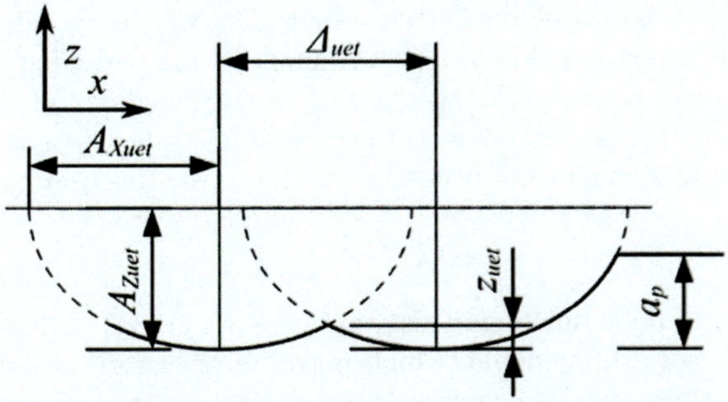

Fig. 4. Illustration of the texture generated by the elliptical movement, side view.

Complete Model. From Eq. (5) and (9), it is possible to define the ideal cutting depth considering f and v_c:

$$a_p = h + z_{uet}. \tag{10}$$

Random vibrations are also added, modeled as a sinusoid along x, with frequency f_z, amplitude A_z and phase ϕ_z:

$$z_V(x) = A_z \sin\left(\frac{2\pi f_z x}{v_c} + \phi_z\right). \tag{11}$$

From Eq. 10, it is possible to define the tool's line constant with origin in $y = 0$:

$$\begin{aligned} B_1 &= -\sqrt{r^2 - y_1^2} + r - a_p + \tan\alpha_1(y_c + y_1) \\ B_2 &= -\sqrt{r^2 - y_2^2} + r - a_p - \tan\alpha_2(y_c + y_2) \end{aligned} \tag{12}$$

And the tool's shape function $z_T(y) \le 0$ can be defined in the interval $y \in [0, f]$:

$$z_T(y) = \begin{cases} -\tan\alpha_1 y + B_1, & \text{if } 0 \le y \le y_1 \\ -\sqrt{r^2 - (y - y_c)^2} + r - a_p, & \text{if } y_1 < y \le y_2 \\ \tan\alpha_2 y + B_2, & \text{if } y_2 < y \le f \end{cases} \tag{13}$$

Similarly, the elliptical texture's shape function $z_U(x) \ge 0$ can be defined in $x \in \left[-\frac{\Delta_{uet}}{2}, \frac{\Delta_{uet}}{2}\right]$:

$$z_U(x) = A_{Zuet}\left[1 - \sqrt{1 - \left(\frac{x}{A_{Xuet}}\right)^2}\right]. \tag{14}$$

Depth in function of position $z(x, y)$ is, therefore, calculated by:

$$z(x, y) = z_T(y) + z_U(x) + z_V(x). \tag{15}$$

However, $z(x, y)$ is only defined in a rectangle $f \times \Delta_{uet}$. To be able to define it in the entire desired area, it is possible to take advantage of the periodicity of the generated patterns. That may be done, for example, with a function analogous to the operator *modulo*, which displays the property $0 \leq (I \bmod N) < N$, with $I, N \in \mathbb{Z}$. More generally, and applicable to real numbers, it is possible to define the function $N(\chi, \delta)$:

$$N(\chi, \delta) = \left\lfloor \frac{\chi}{\delta} \right\rfloor, \tag{16}$$

in which $\lfloor \cdot \rfloor$ is the floor function. This way, $[\chi - N(\chi, \delta)\delta] \in [0, \delta)$. However, the floor function is not differentiable, which makes it necessary to use a differentiable approximation. For this implementation, it was used a continuous approximation based on an approximation of the sawtooth function [13]:

$$\text{sfloor}(w) = w - \frac{\pi + \left[1 - 2\arccos\left((1-\varepsilon_f)\sin\left(\pi\frac{2w-1}{2}\right)\right)\right] 2\arctan\left(\frac{\sin(\pi w)}{\varepsilon_f}\right)}{2\pi}, \tag{17}$$

in which ε_f is a constant such that $\varepsilon_f \to 0$ approximates sfloor(w) to the floor function. Such approximation, however, features two problems: sfloor$(0) = -0.5$ and, when $w \to 0$, sfloor$(w) < 0$. That was worked around by adding a constant ε_y to w when necessary and by using a continuous approximation of the absolute value function:

$$\text{sabs}(w) = \sqrt{w^2 + \varepsilon_a}, \tag{18}$$

in which ε_a is a constant such that $\varepsilon_a \to 0$ approximates sabs(w) to the absolute value function.

In order to better simulate the rotational movement present in turning, the calculations may be done based on the distance traveled by the tool by using the cylinder's perimeter, considering the analyzed cutting line, instead of considering only the simulated area. Therefore, having obtained such line using the function $N_y(y) = \text{sabs}(N(y + \varepsilon_y, f))$ (that is, obtaining the greatest multiple of f which is smaller than y), the distance traveled may be calculated by:

$$x_p(x, y) = N_y(y)\pi D + x, \tag{19}$$

in which D is the cylinder's diameter. Consequently, Eq. (15) may be rewritten as:

$$\begin{aligned} z(x, y) = {} & z_T(y - N_y(y)f) + z_U\left(x_p(x, y) - \left[N_x(x, y) + 0.5\right]\Delta_{uet}\right) \\ & + z_V\left(x_p(x, y)\right) \end{aligned} \tag{20}$$

with $N_x(x, y) = \text{sabs}(N(x_p(x, y), \Delta_{uet}))$. Equation (20) is, therefore, continuous and differentiable, with continuity class at least C^1.

2.2 Mathematical Programming

A discrete environment was used for the optimization process, with $W \times H$ points, spaced 1 μm vertically and horizontally. From that, it is possible to calculate the area's average roughness by:

$$S_a(f, v_c) = \frac{1}{WH}\sum_{x=1}^{W}\sum_{y=1}^{H} \text{sabs}(z(x, y) - z_{med}), \tag{21}$$

in which z_{med} is the height of the mean plane, calculated as the ratio between the surface's volume with plane $z = 0$ (V_0) and the area projected by the surface over that plane (A_0). For the surface area, a topography area calculation method described by Jenness [14] was used: the environment is subdivided into groups of 3×3 points forming 8 three-dimensional triangles per group (Fig. 5). Their areas are then summed and divided by the area of the two-dimensional projection of the surface, $A_0 = (W - 1)(H - 1)$, obtaining the relative surface area $A(f, v_c)$..

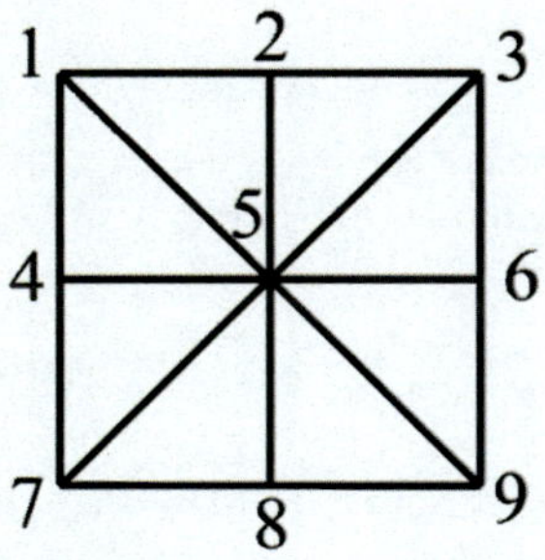

Fig. 5. Illustration of the method used for calculating the surface area, depicting the disposition of the points and the triangles they generate.

Considering the requirements of dental implants, the optimization problem to be solved consists on maximizing surface area with a roughness constraint (maximum roughness $(S_a)_{max}$) in function of the cutting parameters f and v_c, that is:

$$\begin{array}{ll} \max & A(f, v_c) \\ \text{subject to} & S_a(f, v_c) \leq (S_a)_{max} \end{array} \tag{22}$$

It would be ideal, however, for $S_a(f, v_c)$ to be as close as possible to $(S_a)_{max}$, or at least for it to be greater than a minimum roughness, that is, $S_a(f, v_c) \geq (S_a)_{min}$. Figure 6 presents plots of S_a and A which illustrate the behavior of such functions. It is notable that both of them are mostly increasing functions, so their optimization may be done by using linear approximations (sequential linear programming).

By defining the vector $\mathbf{P}_o^K = \left[f \; v_c\right]^T$ at each iteration K it is possible to obtain a parametric straight line based on $A\left(\mathbf{P}_o^K\right)$:

$$\mathbf{P} = \mathbf{P}_o^K + \lambda \widehat{\nabla} A\left(\mathbf{P}_o^K\right), \tag{23}$$

in which $\widehat{\nabla}\cdot$ represents a unitary gradient vector and λ is the line's parameter. One can also define a plane based on $S_a\left(\mathbf{P}_o^K\right)$ which generates an approximation of the constraint function's optimal set of points considering the imposed maximum value, that is:

$$\widehat{\nabla} S_a\left(\mathbf{P}_o^K\right) \cdot \left(\mathbf{P} - \mathbf{P}_o^K\right) + S_a\left(\mathbf{P}_o^K\right) = (S_a)_{max}. \tag{24}$$

The next estimate of the set of cutting parameters is, therefore, the intersection between that line and that plane, which can be obtained analytically: considering the

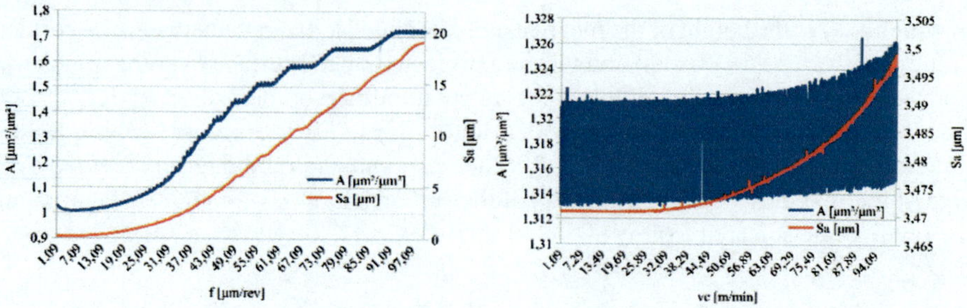

(a) (b)

Fig. 6. Plots of S_a and A in function of cutting parameters: (a) $f \in [1, 100]$ µm/rev with $v_c = 50$ m/min; (b) $v_c \in [1, 100]$ m/min with $f = 40$ µm/rev.

estimate of the next point as $\mathbf{P}_o^{K+1} = \mathbf{P}_o^K + \lambda^{K+1}\widehat{\nabla}A\left(\mathbf{P}_o^K\right)$, it is necessary that

$$\widehat{\nabla}S_a\left(\mathbf{P}_o^K\right)\cdot\left(\mathbf{P}_o^{K+1} - \mathbf{P}_o^K\right) = \widehat{\nabla}S_a\left(\mathbf{P}_o^K\right)\cdot\left[\lambda^{K+1}\widehat{\nabla}A\left(\mathbf{P}_o^K\right)\right] = (S_a)_{max} - S_a\left(\mathbf{P}_o^K\right), \tag{25}$$

therefore:

$$\lambda^{K+1} = \frac{(S_a)_{\max} - S_a\left(\mathbf{P}_o^K\right)}{\widehat{\nabla}S_a\left(\mathbf{P}_o^K\right)\cdot\widehat{\nabla}A\left(\mathbf{P}_o^K\right)}, \tag{26}$$

from which it is possible to obtain the next point $\mathbf{P}_o^{K+1}$, repeating the process until the end condition is reached. For this case, the criterion used was:

$$\left|1 - \frac{A^K}{A^{K-1}}\right| \leq 10^{-6} \tag{27}$$

The method also supports defining upper and lower bounds for the optimization variables. In case a variable $p_i \in \mathbf{P}$, $i = 1, \ldots, n_P$ (n_P being the number of dimensions of $\mathbf{P}$) exceeds the lower bound $(p_i)_{min}$ or the upper bound $(p_i)_{max}$, a new λ^{K+1} is calculated by:

$$\lambda^{K+1} = \min\left\{\lambda^{K+1}, \eta_\lambda \frac{(p_i)_{lim} - p_i^K}{\left(\widehat{\nabla}A\left(\mathbf{P}_o^K\right)\right)_i}\right\}, i = 1, \ldots, n_P \tag{28}$$

in which $(p_i)_{lim}$ is the exceeded bound and η_λ is a dampening constant. For this case, $\eta_\lambda = 0.5$ was used.

2.3 Implementation

To evaluate the described model and the optimization method, they were implemented in C++, and the source code was made publicly available [15]. The SFML library was used to display the simulated topography, the YAML standard was used to define configuration files (using the library yaml-cpp), the OpenMP library was used to accelerate the calculations using parallel programming [16] and Python scripts were also used to aid in visualizing the results.

2.4 Analysis of Results

Experimental analyses have yet not been done, and the model still needs to be modified to be more in line with models already present in the literature and experimentally validated, such as the ones described by Liu et al. [12]. Therefore, the present analysis consists on comparing results based on resolution (using different values of W and H) and considering the exact version of the model (that is, without using continuous approximations). Table 1 displays the parameters used, with $W = H = 100k$, $k = 1, 2, ..., 8$, $(S_a)_{max} = 1.5\,\mu\text{m}$, $f \in [1, 100]\,\mu\text{m}$ and $v_c \in [40, 100]$ m/min.

Table 1. Parameters used to analyze the results. f^0 and v_c^0 are the initial values of the cutting parameters.

Parameter	Value	Parameter	Value	Parameter	Value
ε_f	10^{-10}	α_1	75°	A_{Xuet}	200 µm
ε_a	10^{-15}	α_2	75°	A_{Zuet}	100 µm
ε_y	10^{-3}	r	10 µm	f_z	3×10^9 Hz
f^0	20 µm/rev	D	120 µm	A_z	0.1 µm
v_c^0	60 m/min	f_{uet}	20000 Hz	ϕ_z	0°

3 Results

Table 2 displays the results of the optimizations, and Fig. 7 displays an example of the resulting simulated topography. It is noticeable that the resulting values of f and a_p displayed low variability, and the values of A and S_a displayed even lower variability. On the other hand, v_c varied greatly. The plots in Fig. 6 illustrate the origin of that matter: the feed rate exhibits greater influence over surface area and roughness, however, in general, varying f does not vary them greatly. Cutting velocity, on the other hand, although it contributes less to those two values, has derivatives of greater intensity, making the optimization process less stable. However, as it does not influence the results as much (at least, considering the parameters used), the choice of cutting velocity is not as important as the choice of feed rate and cutting depth.

The use of parallel programming contributed to accelerate the optimization process. By using $W \times H$ points, it is necessary to do $3 \times W \times H$ calculations related to position (calculating $z(x, y)$, $\frac{\partial z(x,y)}{\partial f}$ and $\frac{\partial z(x,y)}{\partial v_c}$) followed by 7 reduction calculations, that is,

Table 2. Results of the optimization process using different resolutions.

W [μm]	f [μm/rev]	ap [μm]	vc [m/min]	A [μm²/μm²]	Sa [μm]	Iterations
100	18.7503	7.12863	48.3265	1.26773	1.49999	1018
200	18.8253	7.49617	59.5587	1.25283	1.50001	90
300	18.8930	7.60258	59.9634	1.25459	1.50001	406
400	18.7950	7.65594	66.8489	1.26272	1.49999	539
500	18.6390	8.45620	95.1740	1.25466	1.49999	175
600	18.8661	7.40710	53.6418	1.26307	1.50000	289
700	18.8178	7.54101	61.6465	1.25623	1.50009	212
800	18.8410	7.55841	61.0871	1.26199	1.50000	256
Average	18.803	7.606	63.3	1.25923	1.50001	373
Standard deviation	0.0793	0.380	14.0	0.00533	0.0000334	295

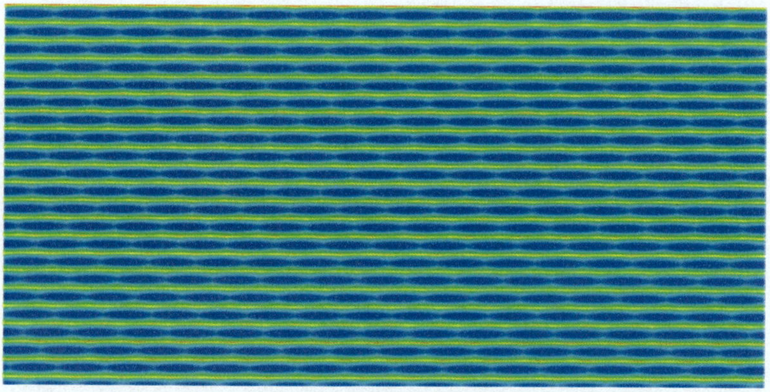

Fig. 7. Detail of the simulated topography for $W = 800$.

using those points to obtain a single value (A, S_a, z_{med}, $\frac{\partial A}{\partial f}$, $\frac{\partial S_a}{\partial f}$, $\frac{\partial A}{\partial v_c}$, and $\frac{\partial S_a}{\partial v_c}$), therefore, the use of multiple processing cores, along with the fact that such calculations are trivially parallelizable, contributes to considerably accelerate that process. However, due to the imprecision inherent to the floating point representation [17] and the OpenMP standard not necessarily guaranteeing a deterministic order of beginning and ending of the concurrent processes [18], small distortions in the calculation results occur during optimization, making it non-deterministic. Although that makes it harder to exactly reproduce results, it also allows one to obtain different results simply by executing the program again.

From those results, there is no apparent relationship between the number of iterations and the resolution used. Although there is an apparent tendency for smaller resolutions to need more iterations, such number may vary from one execution to another due to the effect generated by the use of parallel programming. However, one must also consider that, by increasing resolution, the time per iteration is also increased, therefore, the number of iterations by itself is also not an indicator of solution speed.

Table 3 displays the results obtained by using the cutting parameters obtained from the optimization process in order to recalculate the values of area and roughness without using the continuous approximations. As the results show, the obtained values are close to the ones calculated using the approximations, especially when considering the average and standard deviation. The differences between the obtained values (ΔA and ΔS_a) also show that, with the "exact" area being slightly greater and the roughness displaying certain variability in that regard, that is, varying between being greater or less than the one obtained using the approximations.

Table 3. Results obtained without using the continuous approximations.

W [μm]	A [μm²/μm²]	Sa [μm]	ΔA [μm²/μm²]	ΔSa [μm]
100	1.26759	1.49998	0.000132161	-7.91×10^{-5}
200	1.25263	1.50009	0.000196344	1.33×10^{-5}
300	1.25441	1.50001	0.000176282	-6.88×10^{-6}
400	1.26270	1.50000	0.000017237	-2.99×10^{-6}
500	1.25460	1.49999	0.000059082	-7.70×10^{-6}
600	1.26284	1.50000	0.000236587	6.75×10^{-8}
700	1.25594	1.50010	0.000282801	-3.53×10^{-6}
800	1.26172	1.50000	0.000267872	-3.00×10^{-6}
Average	1.25905	1.5000212	0.0001710	-1.12×10^{-5}
Standard deviation	0.00534	0.0000464	0.0000960	2.82×10^{-5}

4 Conclusions

The use of continuous approximations of discontinuous functions can be considered a suitable way of obtaining continuous approximations of the objective and constraint functions. By using them, it was possible to calculate the gradient of those functions and, with that, to do a gradient-based optimization capable of obtaining adequate cutting parameters considering the requirements of the surfaces of dental implants. The calculated areas and roughnesses were also shown to be close to the ones calculated without the use of such approximations, validating their use.

Regarding the mathematical model, the results were also shown to be practically independent of resolution. However, the model does not encompass matters which could influence the calculation of area and roughness when greater lengths of cut are considered, such as lower frequency vibrations and tool wear. That, therefore, must be analyzed again after the developed model is improved.

For future works, the mathematical model will be better aligned to models present in the literature, such as the one developed by Liu et al. [12], and will also undergo experimental validation, as a way to analyze the viability of using that optimization method to texture dental implants and for other possible applications.

Acknowledgements. The authors would like to thank the Conselho Nacional de Desenvolvimento Científico e Tecnológico — CNPq and the Ministério da Ciência, Tecnologia e Inovações — MCTI for financing this research (process 140441/2022-0).

References

1. Resnik, R.: Misch's Contemporary Implant Dentistry, 4th edn. Elsevier, Canada (2021)
2. Smeets, R., et al.: Impact of dental implant surface modifications on osseointegration. Biomed. Res. Int. **2016**, 1–16 (2016)
3. Kligman, S., et al.: The impact of dental implant surface modifications on osseointegration and biofilm formation. J. Clin. Med. **10**(8), 1641 (2021)
4. Le Guéhennec, L., Soueidan, A., Layrolle, P., Amouriq, Y.: Surface treatments of titanium dental implants for rapid osseointegration. Dent. Mater. **23**(7), 844–854 (2007)
5. Gittens, R.A., et al.: A review on the wettability of dental implant surfaces II: biological and clinical aspects. Acta Biomater. **10**(7), 2907–2918 (2014)
6. Toumeh, T., Sghaireen, M., Ganji, K., Mathew, M., Nagy, A., Rao, K.: A review on biofilm and biomaterials: prosthodontics and periodontics perspective. J. Int. Oral Health **12**(6), 504 (2020)
7. Albrektsson, T., Wennerberg, A.: The impact of oral implants - past and future, 1966–2042. J. Can. Dent. Assoc. **71**(5), 327 (2005)
8. Coelho, P.G., et al.: Basic research methods and current trends of dental implant surfaces. J. Biomed. Mater. Res. B Appl. Biomater. **88B**(2), 579–596 (2009)
9. Brehl, D., Dow, T.: Review of vibration-assisted machining. Precis. Eng. **32**(3), 153–172 (2008)
10. Saptaji, K., A. G, M., Azhari, A.: Machining of biocompatible materials: a review. Int. J. Adv. Manuf. Technol. **97**, 2255–2292 (2018)
11. Sajjady, S.A., Lotfi, M., Amini, S., Toutounchi, H., Bagheri-Bami, A.: Improving the surface energy of titanium implants by the creation of hierarchical textures on the surface via three-dimensional elliptical vibration turning for enhanced osseointegration. Proc. Inst. Mech. Eng. [H] **233**(12), 1226–1236 (2019)
12. Liu, X., Wu, D., Zhang, J., Hu, X., Cui, P.: Analysis of surface texturing in radial ultrasonic vibration-assisted turning. J. Mater. Process. Technol. **267**, 186–195 (2019)
13. Helbling, C.: Smooth floor function. Mathematics Stack Exchange (2020), https://math.stackexchange.com/q/3850631. Accessed 12 Oct 2022
14. Jenness, J.S.: Calculating landscape surface area from digital elevation models. Wildl. Soc. Bull. **32**(3), 829–839 (2004)
15. de Oliveira, T.L.: textopt - optimization of cutting parameters based on fast topography simulation v0.1.1 [Computer Software] (2022)
16. Dagum, L., Menon, R.: OpenMP: an industry standard API for shared-memory programming. IEEE Comput. Sci. Eng. **5**(1), 46–55 (1998)
17. IEEE: 754–2019 - IEEE Standard for Floating-Point Arithmetic (2019)
18. OpenMP: OpenMP API 5.2 specification (2021)

Author Index

D. de Oliveira et al. (Eds.): COBEF 2023, LNME, pp. 231–232, 2024.
https://doi.org/10.1007/978-3-031-43555-3